The Story of INDIA'S UNICORNS

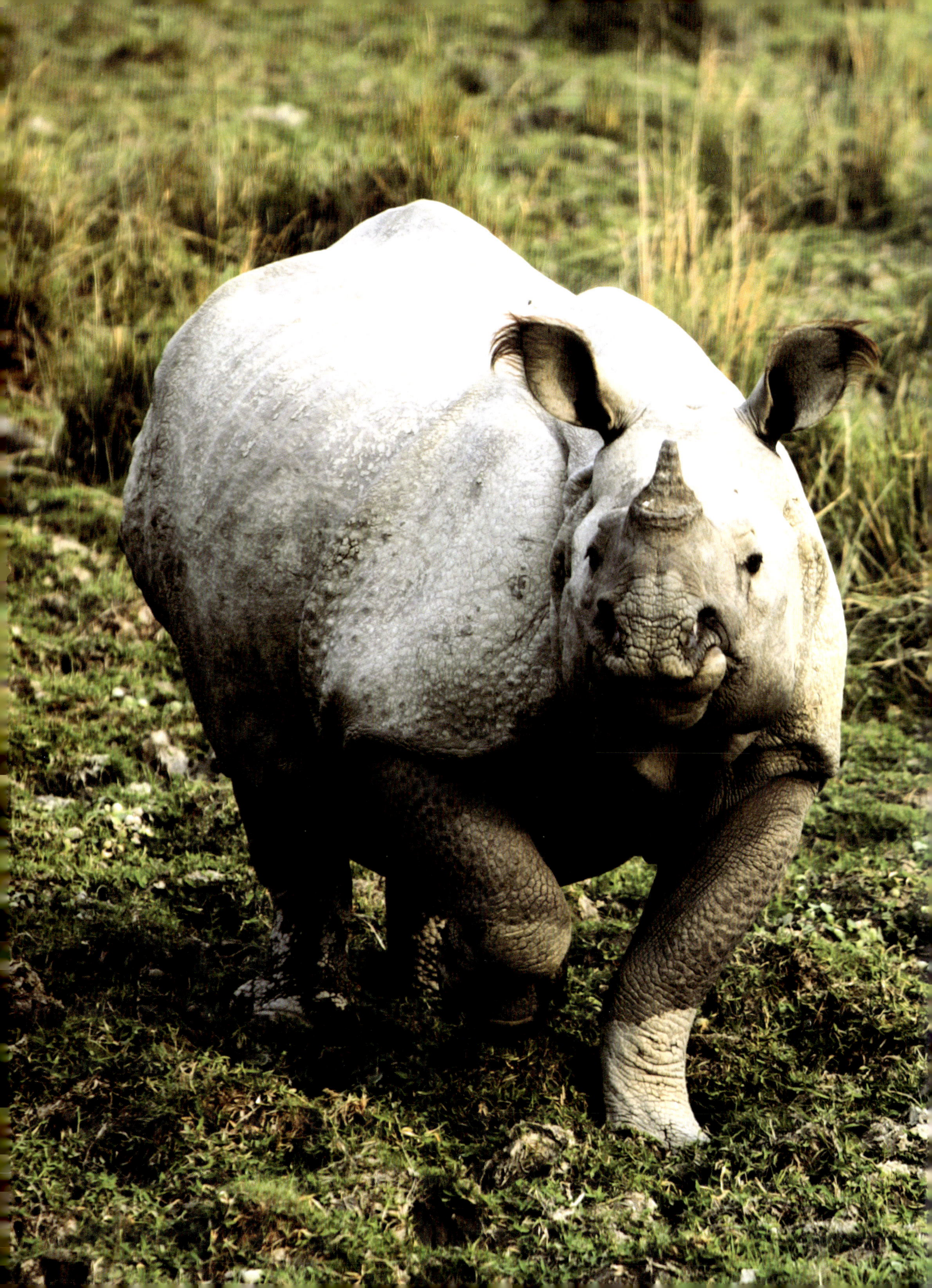

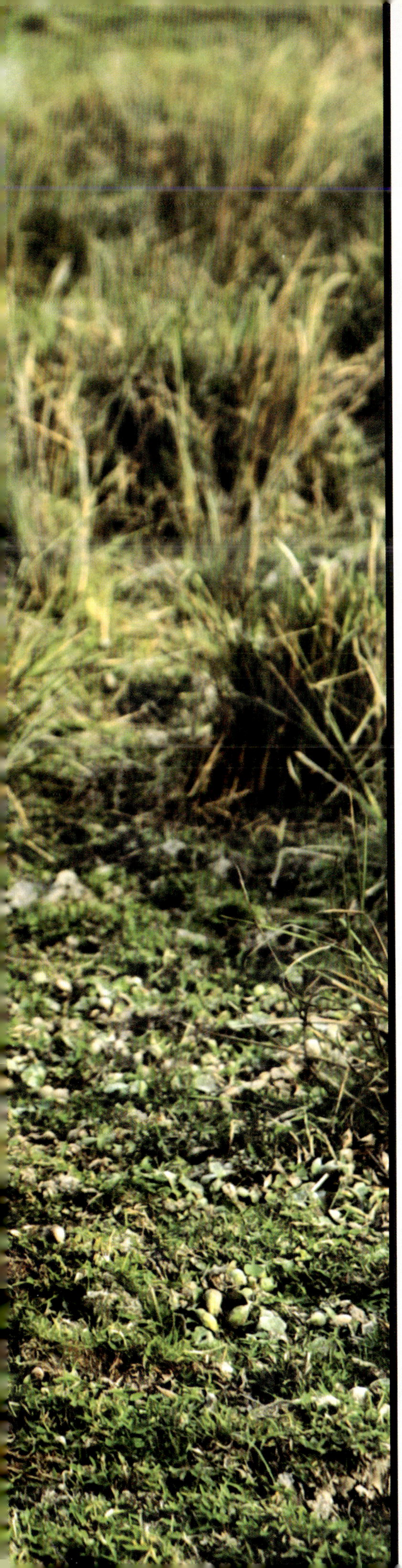

The Story of INDIA'S UNICORNS

Divyabhanusinh,
Asok Kumar Das and Shibani Bose

Mārg

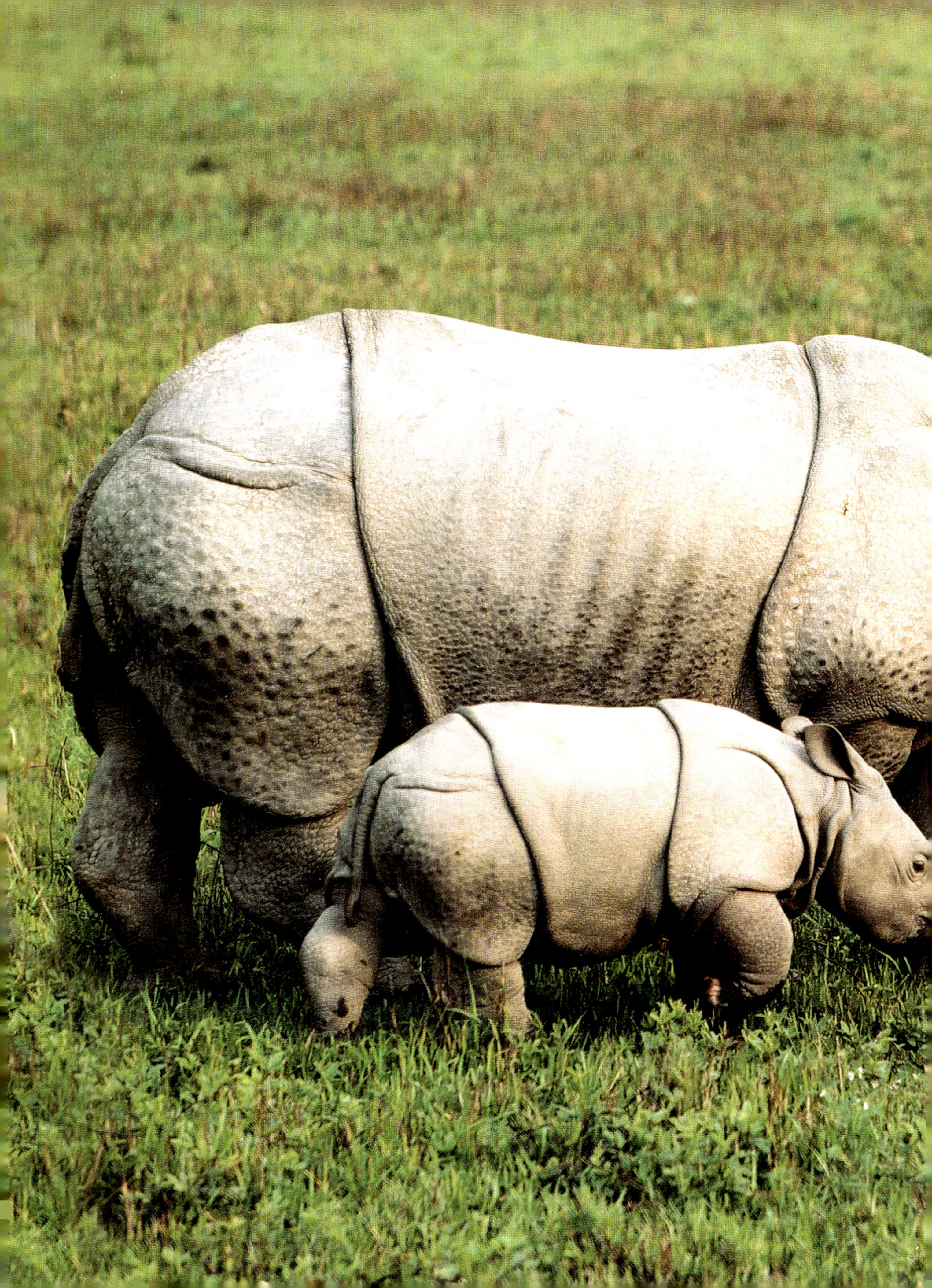

The publication of this book has been made possible by generous support received from
The Corbett Foundation
The Duleep Matthai Nature Conservation Trust (DMNCT)
Hemendra Kothari Foundation
Jagdish Trivedi
Kunj Trivedi
Pirojsha Godrej Foundation

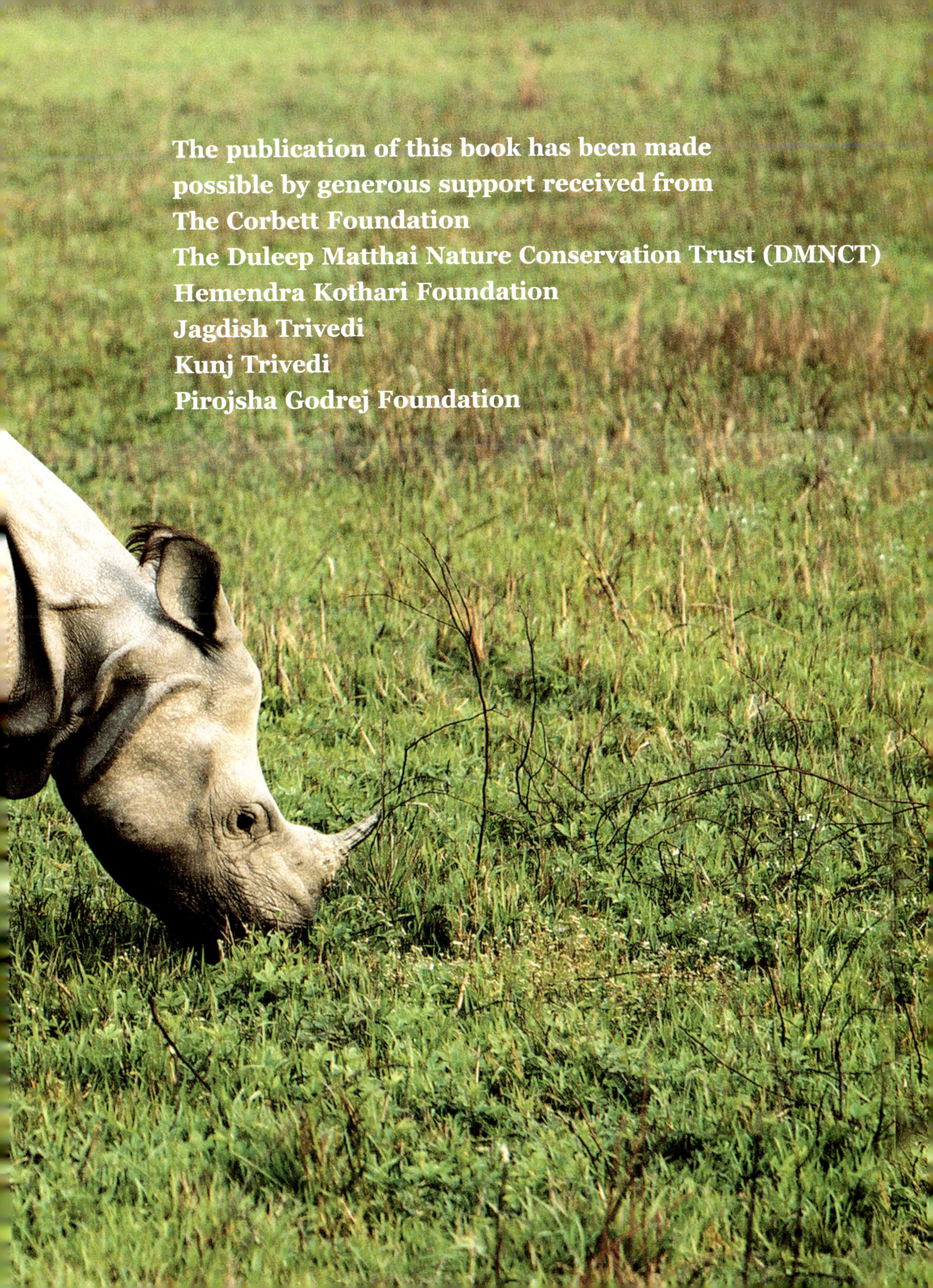

Editors Jyotindra Jain | Naman P. Ahuja
Assistant Editors Jyotsna Nambiar | Mrinalini Vasudevan
Editorial Executive Nisha S. Nair
Text Editor (Consultant) Rivka Israel
Designer Naju Hirani
Senior Production Executive Gautam V. Jadhav
Production Executive Chetan S. More

Price: ₹ 2000.00 / US$ 48.00
ISBN: 978-93-83243-23-5
Library of Congress Catalog Card Number: 2018-305001

The Corbett Foundation will receive 5% of the net sales value of each copy sold

Published by Radhika Sabavala for The Marg Foundation at Army & Navy Building (3rd Floor), 148, M.G. Road, Mumbai 400 001, India.

Processed at The Marg Foundation, Mumbai.

Printed at Silverpoint Press Pvt. Ltd., Navi Mumbai.

Cover ▪ *Front:* Kaziranga.
Photograph: Kakubhai Kothari.
Back: Daimabad bronze.
Courtesy National Museum, New Delhi.

Page 2 ▪ A charging rhino, Kaziranga.
Photograph: Kakubhai Kothari.

Pages 4–5 ▪ Rhino mother and calf, Kaziranga.
Photograph: Kakubhai Kothari.

Page 8 ▪ Face-off: two male rhinos dispute territory, Kaziranga.
Photograph: Kakubhai Kothari.

Page 11 ▪ The victor chases off the vanquished, Kaziranga. The hog deer in the foreground are not disturbed.
Photograph: Kakubhai Kothari.

contents

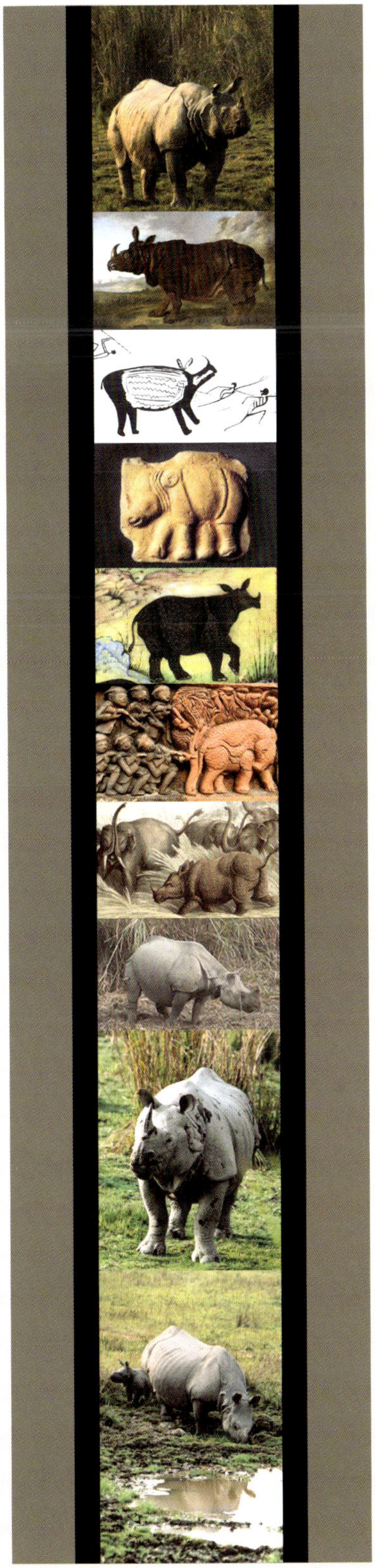

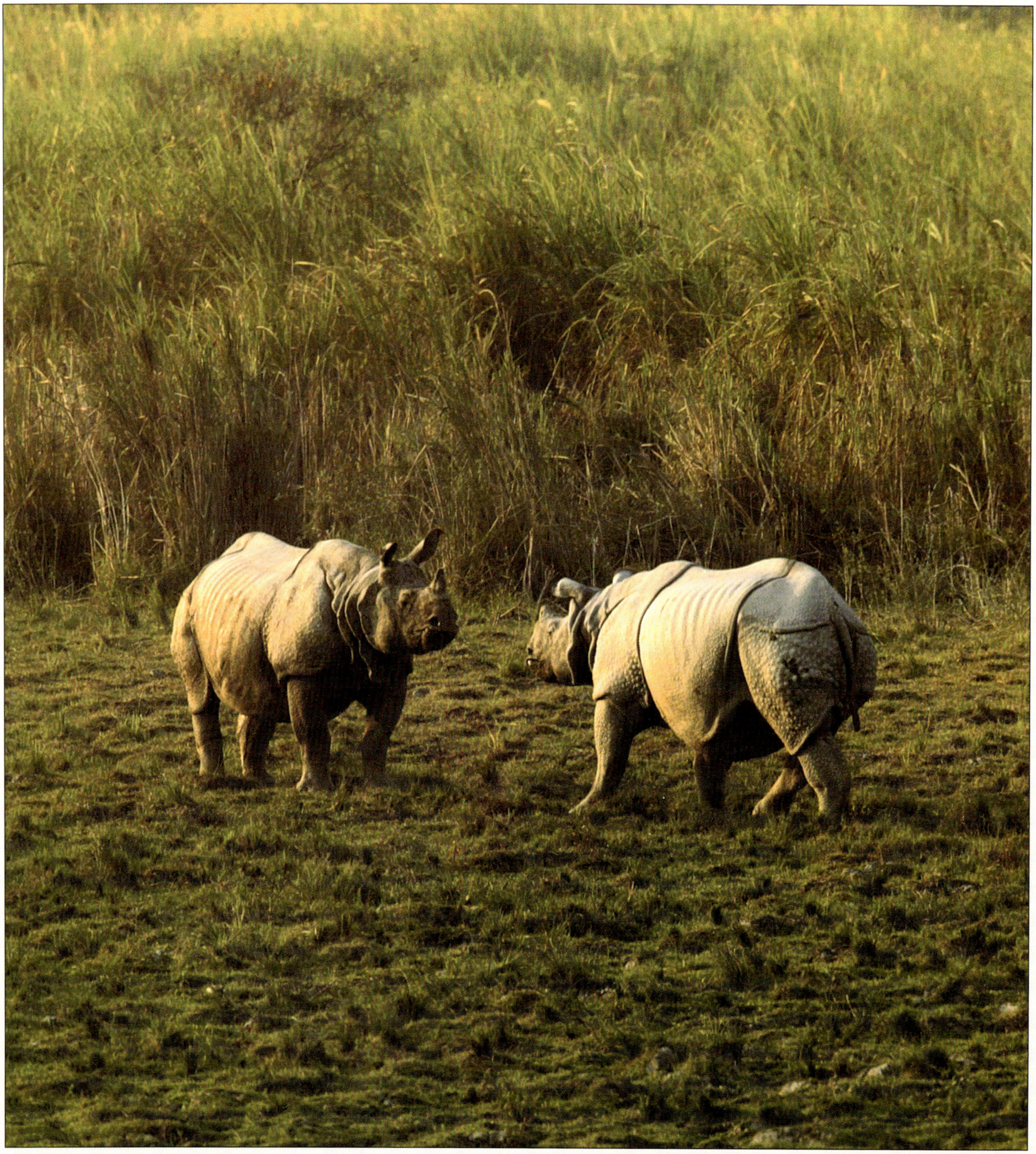

India is blessed with a unique variety of fauna, not to mention other forms of life. If one were to look at its megafauna alone, the country boasts of lion, tiger, leopard, elephant, gaur, wild buffalo and rhinoceros—more specifically the Greater One-horned Rhinoceros.

Some large fauna have fallen by the wayside owing to hunting (though this has now been banned for some time, the threat of poaching continues) as well as increasing human pressures and consequent loss of natural habitats. This has been a continuous process. We have lost the cheetah in the plains of the Indian subcontinent as also the Javan and Sumatran rhinoceroses in their extreme western range in eastern India in historical times. The natural habitats of all species of fauna and flora are under threat as a result of the present and ever-increasing pressure of the growing human population which stands today at 1.32 billion in India. Not only are the megafauna under constant threat, smaller species too, whose fate we are largely unaware of, are in equal danger.

Two books have preceded this one in "The Story" series: one on Asia's lions and the other on Asia's elephants. *The Story of India's Unicorns* is an attempt at recording the history of the animal from the earliest times to the present, as was done in the other two books. It chronicles the travails of the Greater One-horned Rhinoceros which is endemic to the Indian subcontinent. It is entirely up to India, Nepal and possibly Bhutan to ensure that this species survives for all time. The authors hope that their contributions will interest a wider audience in the protection of the animal which at present enjoys a "return" of sorts. What the future holds for India's unicorns remains uncertain.

THE WRITING OF THIS BOOK has taken the authors far and wide to hunt out references to the unicorn over the ages. They have accumulated many debts during their journeys.

Shibani Bose is deeply indebted to Professor Nayanjot Lahiri, her doctoral supervisor at the University of Delhi, and to Professor Mahesh Rangarajan (both now at Ashoka University). Their valued insights have not only enabled her to push her horizons, but have also helped her to understand the various nuances the subject encompasses. She gratefully acknowledges the time spent by Dr G.L. Badam of Deccan College, Pune, who generously shared his expertise.

Asok Kumar Das offers sincere thanks to Dr Michael Brand, Dr Jorrit Britschgi, Dr Debra Diamond, Dr Barbara Karl, Mr Simon Ray, Mr Navin Kumar, Dr Daniel Ehnbom, Ms Rosemary Crill, Mr Rajesh Purohit, Dr Andrew Topsfield, Dr Bikramjit Ram, Ms Holly Shaffer, Ms Ana Kol and Professor Subhash Parihar for their invaluable help.

Mr Dilip Khatau, Chairman of Corbett Foundation, has been a strong supporter of rhino conservation. The CEO of the Foundation Dr Kedar Gore went to great lengths to collect information on poaching, and on rhino numbers and conditions in Assam. At the World Wide Fund for Nature–India all-round support and assistance was forthcoming throughout: Dr G. Areendran prepared the two maps for the book; Ms Dilpreet B. Chhabria of Traffic-India searched for and procured invaluable illustrations; Ms Maisra Malik's assistance in preparing the bibliography was invaluable; while Ms Sunitha Jaykumar promptly, cheerfully and efficiently typed out the manuscript apart from meeting innumerable demands of preparing the manuscript for the publishers. Mr Ravi Singh, an old friend and Secretary General and CEO of WWF–India, was as usual unfailingly generous with books, advice and assistance. In Mumbai, Mr Kakubhai Kothari of Photokina unstintingly offered a choice of his excellent rhinoceros images from Kaziranga which have substantially added to the visual appeal of the book, while K. Somendra Singh photographed several of the images for reproduction here. Dr Kees Rookmaaker's fabulous bibliography of the world's rhinoceroses was a great inspiration, and he was equally forthcoming in providing a rare image of King George V's rhinoceros shoot and photographs of Javan and Sumatran rhinos. Dr Deborah Swallow of the Courtauld Institute, London, promptly arranged permission for the use of two images from the Institute's collections. Mr Kenneth X. Robbins permitted use of images from his collection in the USA, while Maharaja Anant Narain Singh of Banaras sent an invaluable image of the "hauda" with the rhinoceros likeness used for official functions. Dr M.K. Ranjitsinh and Professor Mahesh Rangarajan have been a source of inspiration and support for this effort as they have been earlier. Divyabhanusinh gratefully acknowledges the contributions of all the above.

Each of these benefactors has added considerably to the value of the book. Its shortcomings, of course, are the responsibility of the authors alone.

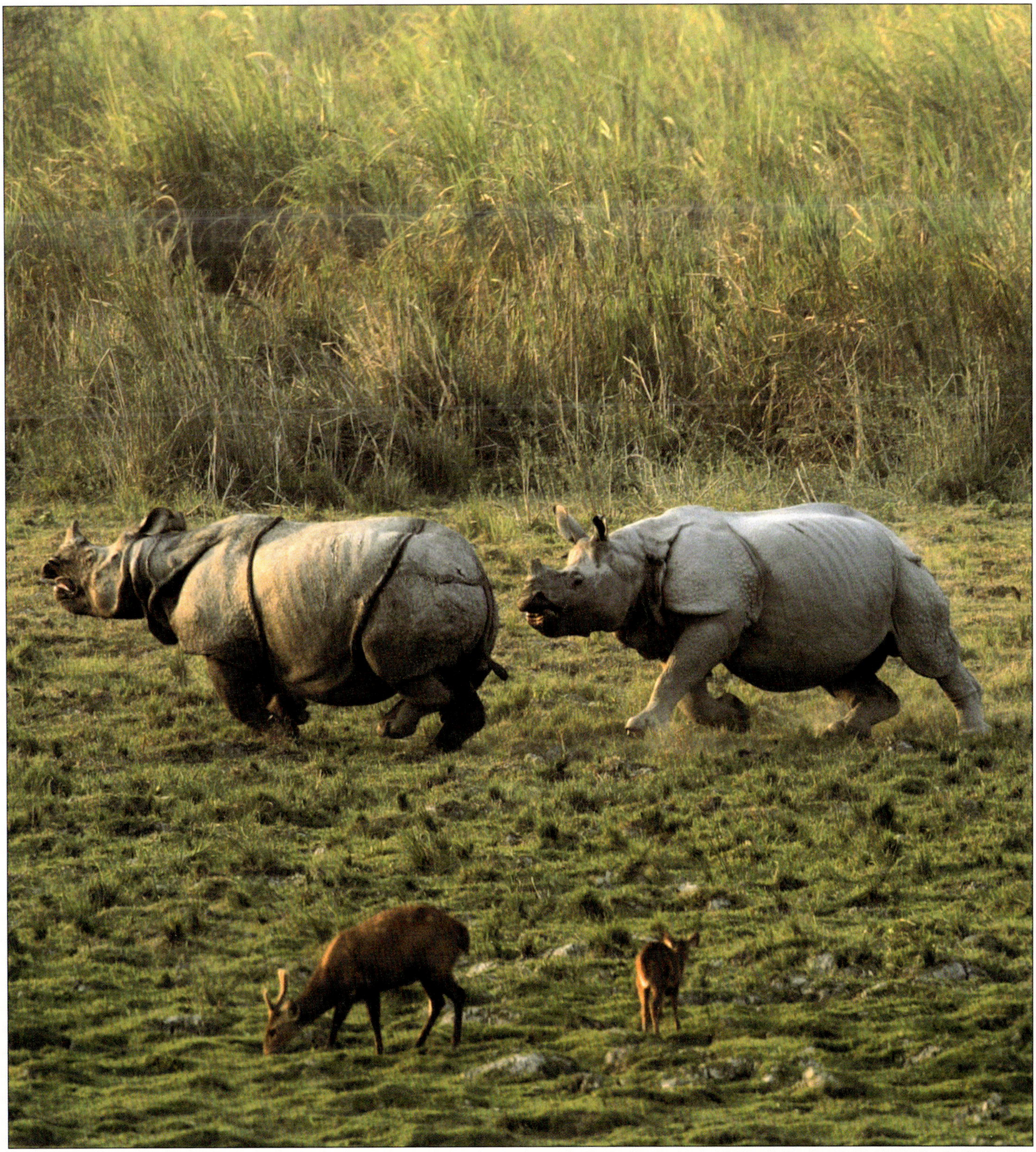

CHAPTER 1

India's Unicorns

Divyabhanusinh

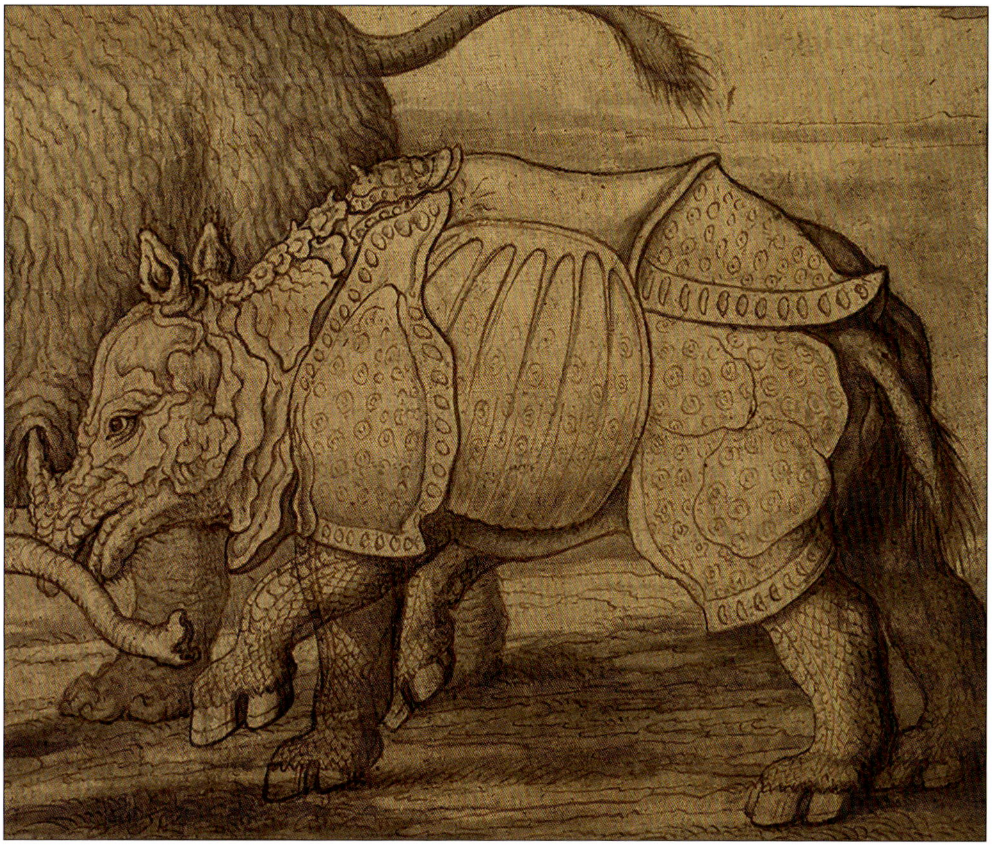

If one were to evoke images of a bygone geological period and imagine animals of that age which are amongst us even today, three megafauna still found on the Indian subcontinent would come to mind: the estuarine or saltwater crocodile found most famously in the Sundarbans mangroves, which is the largest crocodile on earth; Asia's much-loved elephant writ large through history; and the greater one-horned rhinoceros—India's unicorn. The saltwater species of crocodile is arguably the least known to most Indians, while the rhinoceros is more often familiar through pictures and as brand mascot of some products, but it is not met with easily as it has been pushed primarily to the easternmost corner of India. Before taking a closer look at the animal and outlining its treatment in this book, this chapter provides a brief account of rhinoceros evolution, takes note of the other Asian species, and attempts an explanation of how India's species came to be called *Rhinoceros unicornis*.

Evolution

In the mists of geological time which goes back more than 4,000 million years, it is only approximately 65 million years ago, in the Tertiary and Quaternary periods—being the so-called ages of mammals and humans—when the rhinoceros appeared along with other mammals. Their phylogeny—that is, "evolutionary relationships between taxonomic levels, particularly the patterns of lines of descent..."[1]—is a story by itself. For our purposes, I have relied heavily on the works of Eric Dinerstein to survey very briefly the descent of the unicorn and much else besides that follows.

The earliest rhinoceros-like animals appeared in the Eocene epoch of the Tertiary period of the earth's evolution which commenced approximately 54 million years ago and ended about 17.5 million years later. These animals had no horns. Animals that resembled our present-day specimen also made their appearance in the same epoch in Eurasia and North America.

This family, Rhinocerotidae, prospered in the Oligocene epoch which began approximately 36.5 million years ago and lasted about 12.5 million years. During this epoch, Africa still had no Rhinocerotidae. The North American rhinoceroses were first off the block; they grew bigger and grew horns before their brethren in Eurasia. It appears that climate change spelt doom for many groups of rhinoceroses towards the end of the Miocene epoch which began approximately 24 million years ago and lasted about 19 million years. At this time the North American rhinoceroses disappeared. In Asia the group of rhinoceroses to survive was *Dicerorhinines* which gave rise to a form like the Sumatran rhinoceros of today which has changed little since then. The other, more recent lineage, *Rhinocerotines*, diverged in the Pleistocene epoch into the Javan rhinoceros and Indian rhinoceros of today.

In Africa we have two surviving species of rhinoceroses. The black rhinoceros appeared 4 million years ago approximately, in the Pliocene epoch which commenced approximately 5 million years ago and lasted about 3 million years. The white rhinoceros appeared later in the middle Pliocene.[2]

Asian Species

We have noted that the Sumatran rhinoceros was the earliest to appear in Asia and has changed little from its ancestors which walked the earth 40 million years ago. It has two horns and it may be more closely related to its African cousins which sport two horns too. The other two Asian rhinoceroses, the greater one-horned rhinoceros and the Javan rhinoceros, appeared in the middle Pleistocene, the first of the two

▪ A rhino obliges the photographer with a perfect pose, Kaziranga.
Photograph: Morales/age fotostock/Dinodia.

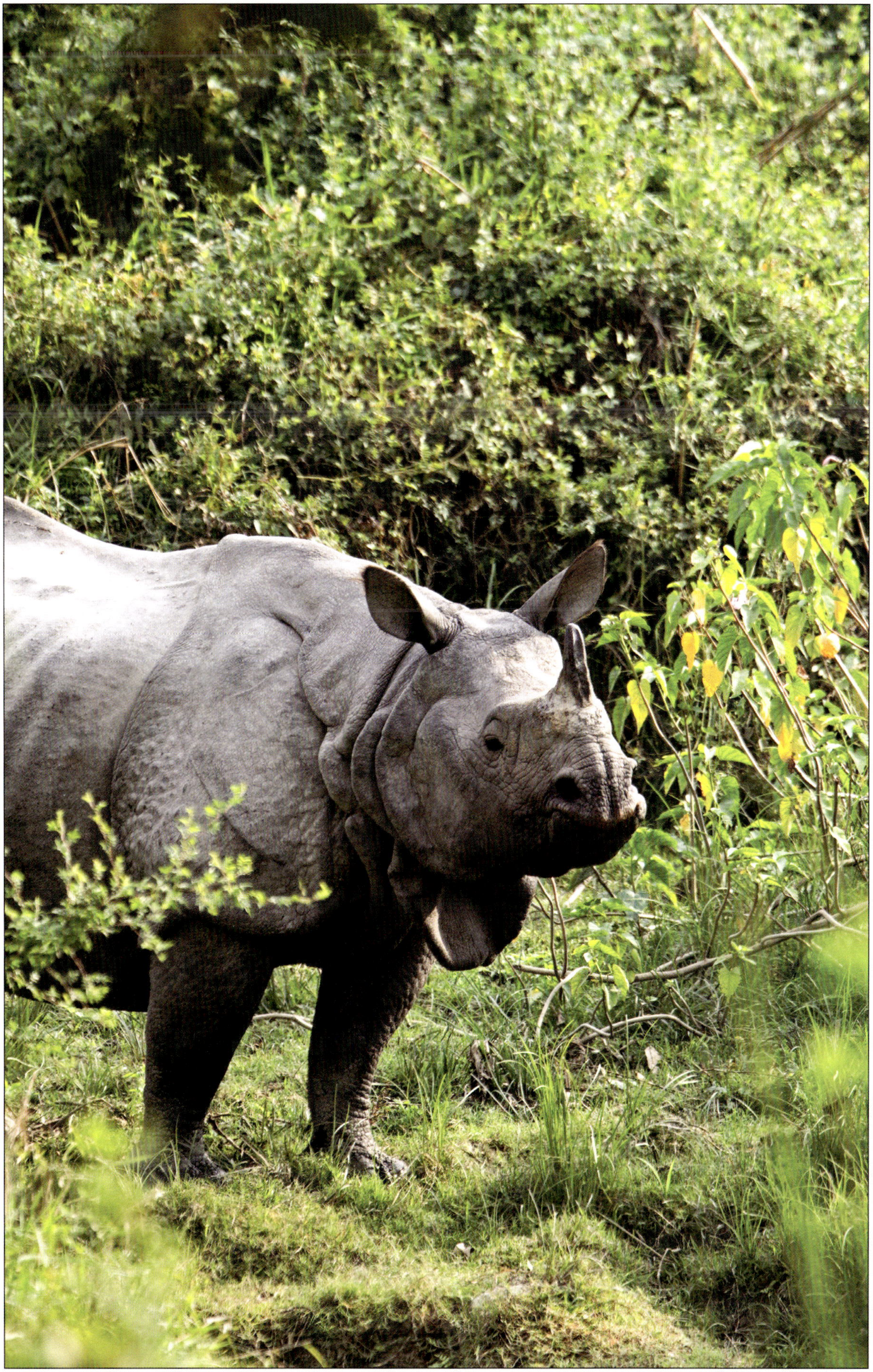

TABLE: RHINOCEROS EVOLUTION

Period	Epoch	Approx. years BP	Rhinoceros Evolution
Tertiary		65 million	
	Palaeocene		---
		54 million	
	Eocene		Rhinoceros-like animals appear in Eurasia and North America — no horns
		36.5 million	
	Oligocene		Rhinocerotidae prospers in North America and in Eurasia — horns present
		24 million	
	Miocene		North American rhinos disappear; Dicerorhinines appears in Asia (two-horned) — ancestor of Sumatran species
		5 million	
	Pliocene		Two-horned rhinos appear in Africa — black rhino appears first; then white rhino; Rhinocerotines appears in Asia (one-horned)
		2 million	
Quaternary	Pleistocene		Rhinocerotines diversifies into Javan and Indian species
		10,000	
	Holocene		
		to present	

epochs of the Quaternary period which commenced some 2 million years ago and lasted till 10,000 years ago approximately.

All the three living rhinoceros species of Asia once inhabited the Indian subcontinent. Thus, there were two unicorns in our midst, the Indian rhinoceros and the Javan rhinoceros, along with the two-horned Sumatran rhino. However, the evidence we have suggests that both the Sumatran and the Javan rhinoceroses were confined to the east and northeast regions of the Indian subcontinent, which was their extreme westernmost range.

With the coming of the British, "scientific" literature came to be published which gives us an idea of the animals' range in India. The earliest of the lot was T.C. Jerdon's magnum opus *The Mammals of India* which was first published in 1867. He recorded that the Javan rhinoceros which he called the "Lesser Indian Rhinoceros" was to be found in the Sundarbans, very few in the tract along the

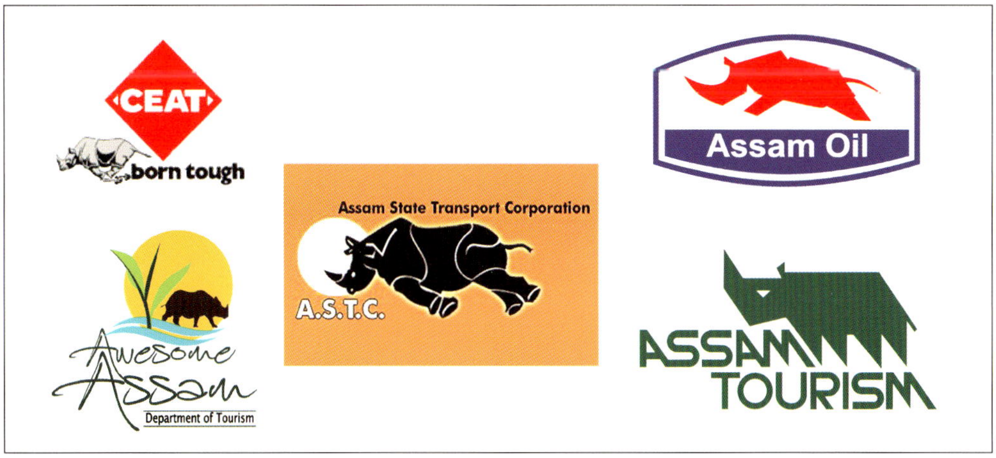

■ Pride of Assam: the people of the state consider the "gainda" as their very special possession. Hence its appearance as symbol on many government and non-government signages. And for decades the "tough" rhino featured in the logo of CEAT tyres.

Mahanadi extending north to Midnapore and on the northern edge of the Rajmahal hills near the river Ganga. He noted that "Several have been killed quite recently within a few miles of Calcutta", and that these animals were found abundantly in Myanmar, and from the Malaysian peninsula to the islands of Java and Borneo.[3] One female Javan rhinoceros lived in the Zoological Garden at Alipore, Calcutta from 1887 until her death in 1892. Before arriving there, she had lived for 10 years in the menagerie of Wajid Ali Shah, the last Nawab of Avadh whose kingdom was annexed by the British.[4] It is not known how the Nawab came by this animal.

Of the Sumatran rhinoceros Jerdon writes that they were shot as high as latitude 23°N near Sandoway (Myanmar) and it was suspected by E. Blyth of the Asiatic Society of Bengal that their range extended to as far north as Assam. One hairy-eared rhinoceros (*Rhinoceros lasiatis*) and one Sumatran rhinoceros (*Rhinoceros sumatrensis*), now recognized as one species going by the name *Dicerorhinus sumatrensis*, lived in the Alipore zoo. The former was there from 1882 and was alive till 1892 while the latter lived there from 1882 to 1887. The provenance of neither is known.[5]

R.A. Sterndale, whose work on the mammals of India and Ceylon followed in 1884, states that the Javan rhinoceros inhabited the Sundarbans, Tipperah (Tripura), the swamps of Garo, Khasi and Naga hills, and Manipur—extending into provinces of China and southwards to Myanmar, the Malaysian peninsula, Sumatra, Java and Borneo. He, however, goes on to note that there were two forms of two-horned rhinoceroses, one being the ear-fringed rhinoceros (presumably what Jerdon called hairy-eared) and the other the Sumatran rhinoceros—a specimen of the former was caught near Chittagong in 1868 and both were found in Tenaserrin provinces while the latter was to be found in Myanmar, Thailand, the Malaysian peninsula and Sumatra.[6] It may be noted that only one species of the Sumatran rhinoceros is recognized today, as stated above.

R. Lydekker's book *The Game Animals of India, Burma, Malaya and Tibet* was first published in 1907 and went through several editions. Even up to its 1924 edition, he still mentions the existence of the Javan and Sumatran rhinoceroses on the Indian subcontinent in its eastern regions.[7] S.H. Prater, the last of the British-period naturalists to write an authoritative work on larger mammals of India, whose book was published for the first time in 1948 a few months after the independence of India and Pakistan, records that the Javan rhinoceros was "now practically exterminated in Assam"[8] which at the time meant all the northeastern states of

▪ The Sumatran rhinoceros has two horns like its African counterparts, and it is the oldest rhinoceros species living. *Courtesy Kees Rookmaaker.*

▪ The Javan rhinoceros is similar in appearance to the greater one-horned rhinoceros but is a much smaller animal. *From Lydekker, 1924.*

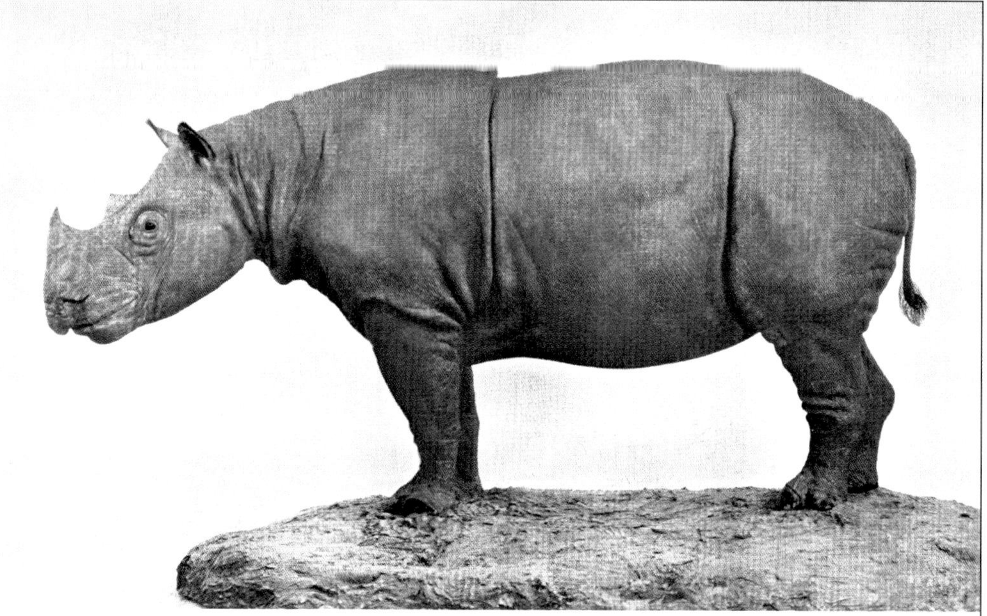

By Courtesy *The Amer. Mus. Nat. Hist., New York.*

The Sumatran Two-horned Rhinoceros (*Rhinoceros sumatrensis* Cuv.).
(*Height at shoulder about 52 in.*).

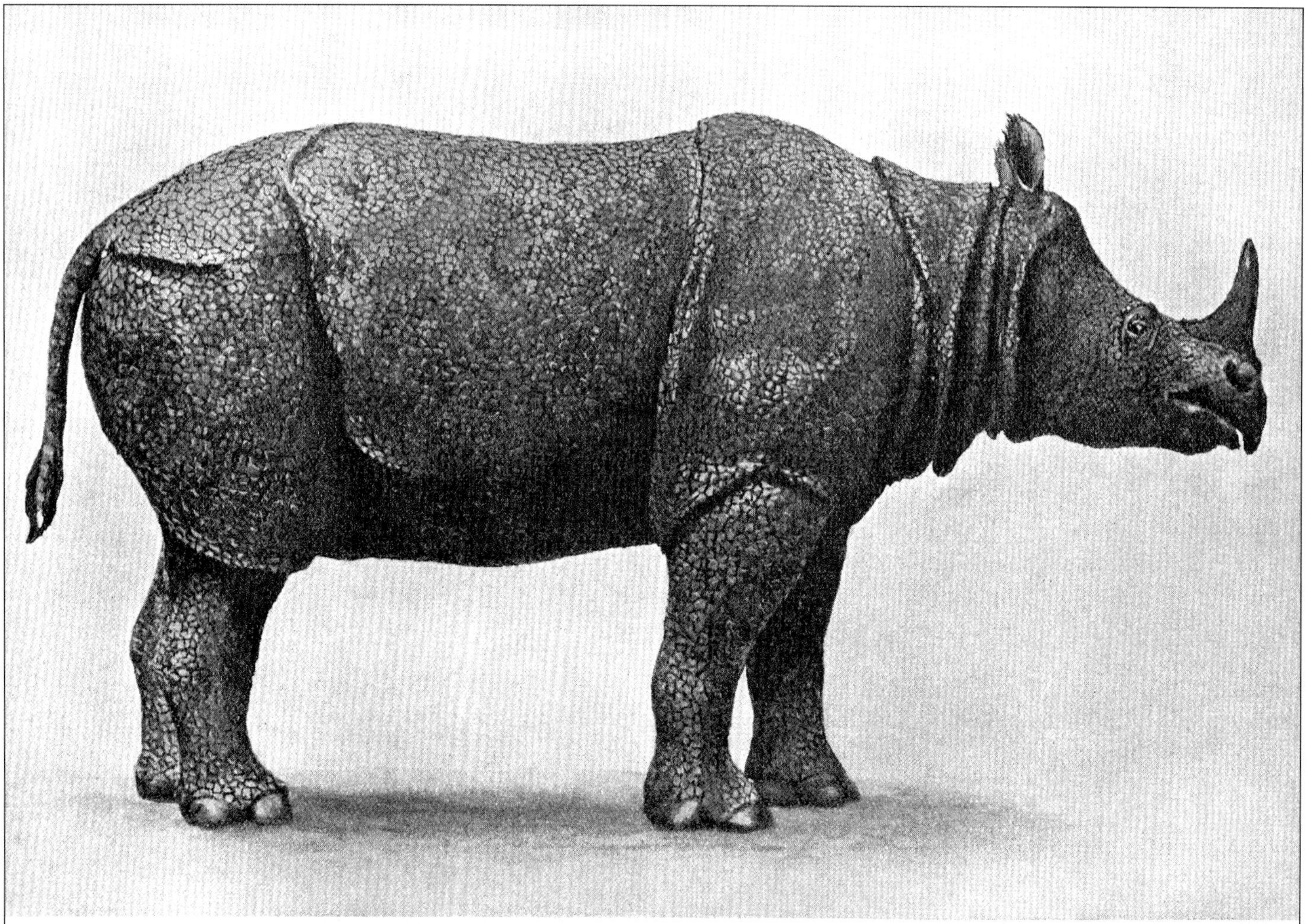

FIG. 1*b*.—Javan Rhinoceros (male).

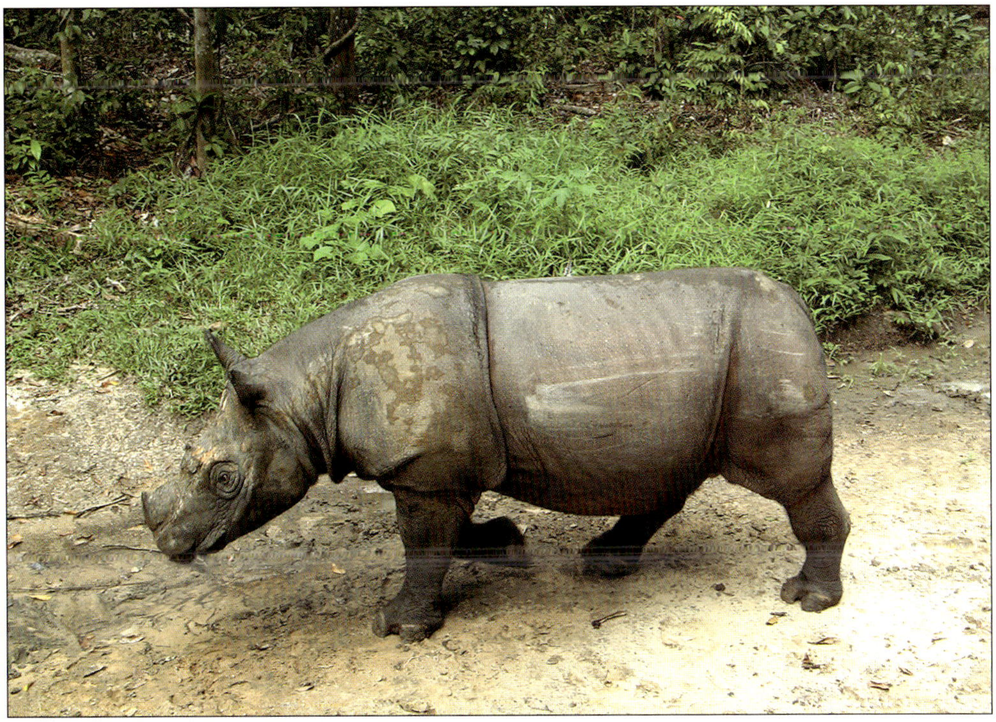

- A Sumatran rhinoceros in the wild. Note the hairy ears.
Courtesy Kees Rookmaaker.

- A Javan rhinoceros in the wild.
Courtesy Kees Rookmaaker.

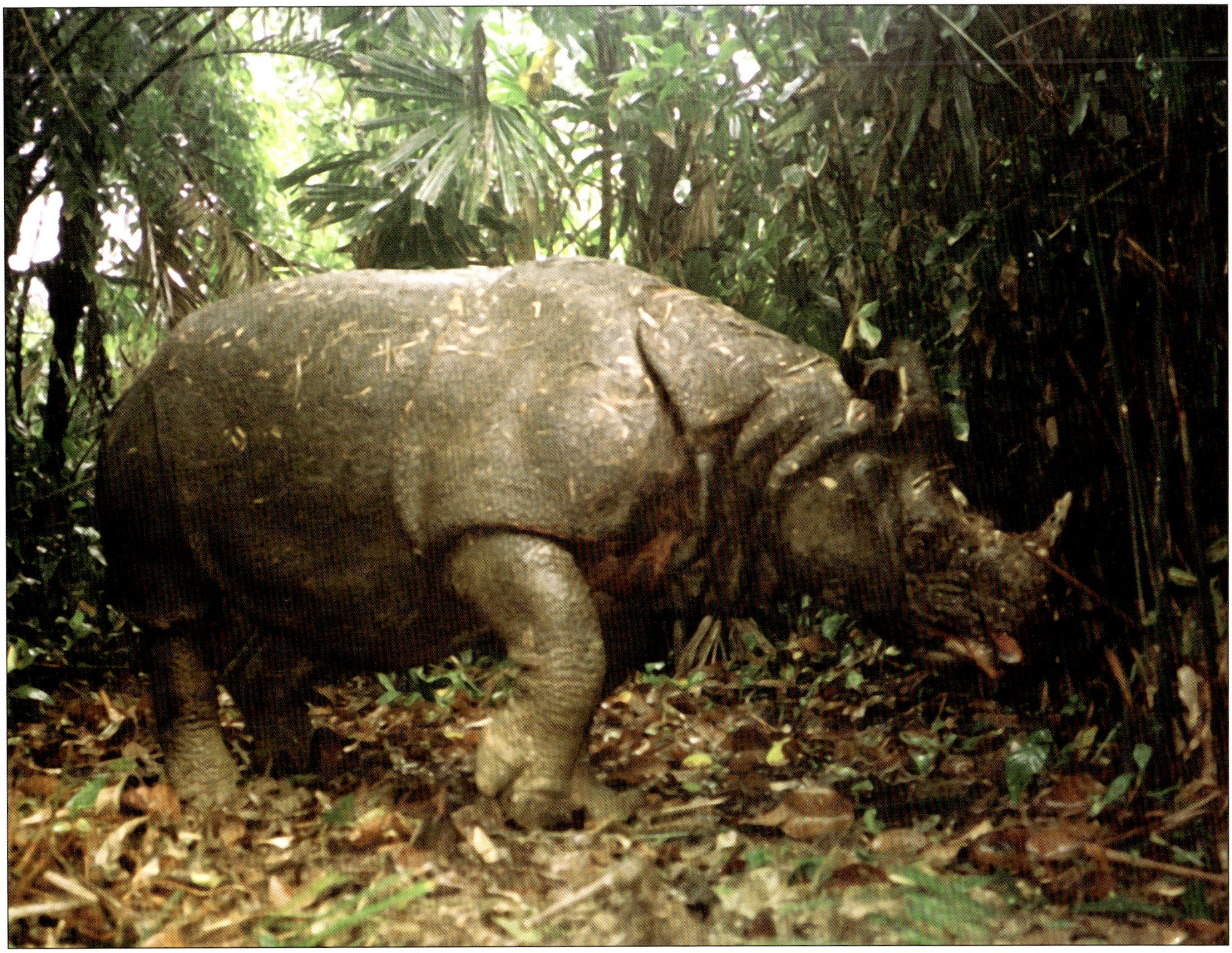

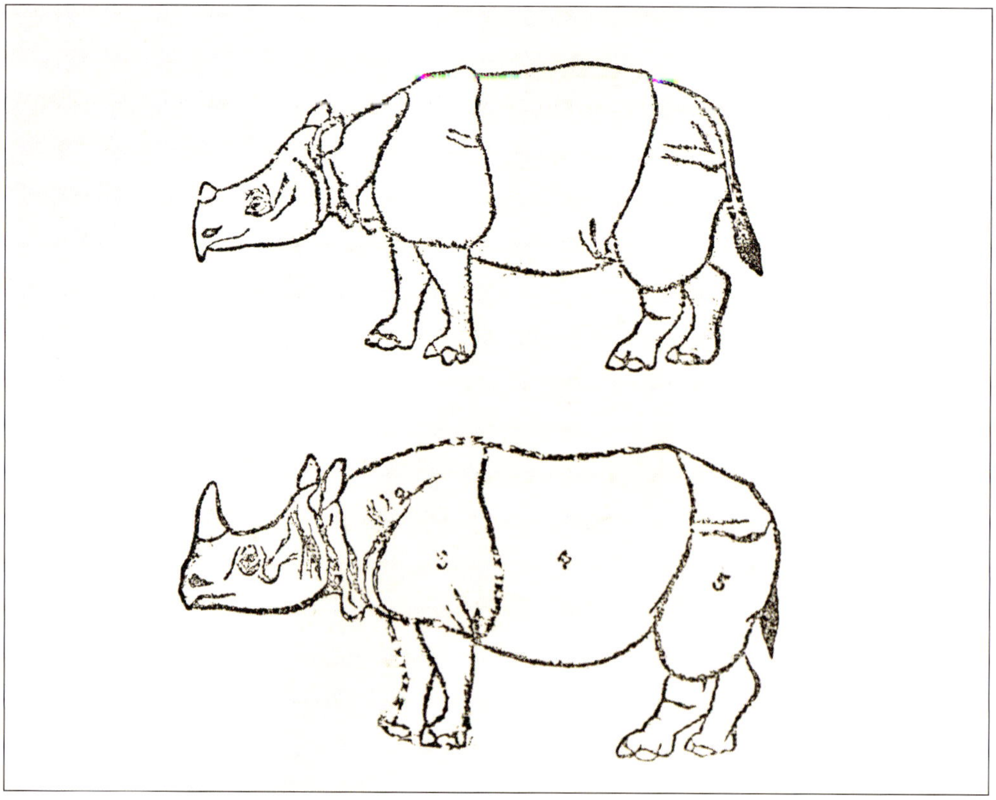

▪ The relative size of the Javan rhinoceros, compared with the greater one-horned rhinoceros.
From Finn, 1929.

today barring Tripura and Manipur. A recent authority has noted that the last Javan rhinoceros lived in the Sundarbans as late as circa 1925.[9]

The Sumatran rhinoceros is the smallest of the three Asian species and is noticeably distinct to the common viewer's eye insofar as it is hairy and has two horns. The Javan rhinoceros, on the other hand, looks similar to the greater one-horned (until recently styled as the Great Indian One-horned) rhinoceros but is a smaller animal. With the departure of these two species from our midst, we are left with only one unicorn which has fared somewhat better than its two other Asian cousins. Kaziranga National Park in Assam has its largest population of some 2,400-plus animals. The Chitwan National Park in Nepal has the second largest population according to an estimation done in 2011, of 605 animals.[10] Additionally, there are scattered groups of small numbers elsewhere in the Terai, West Bengal and Assam. The total wild population of the animal is estimated to be around 3,500 only (from a low of about 200 in 1900).[11] This has led to its being classified as vulnerable on the IUCN (World Conservation Union) Red List and it appears in Appendix I of CITES (Convention on International Trade in Endangered Species) which gives it protection with a ban on trade of the animal itself and its body parts among other things.

Names and Descriptions

The greater one-horned rhinoceros has been written about in literature from ancient times. It is known as *gainda* in Hindi and *gondar* in Bengali, while in Assam, its last stronghold, it is sometimes called *gaur*—the name also given to India's mega bovine *Bos gaurus* prominent in the peninsula. In Hindi in the past the rhino has also been called *gargadan,* a word which is unknown in popular usage today, and which appears to be a corruption of the Arabic word for the animal—*karkadan.* Persian uses the word *kark* or *karg.*

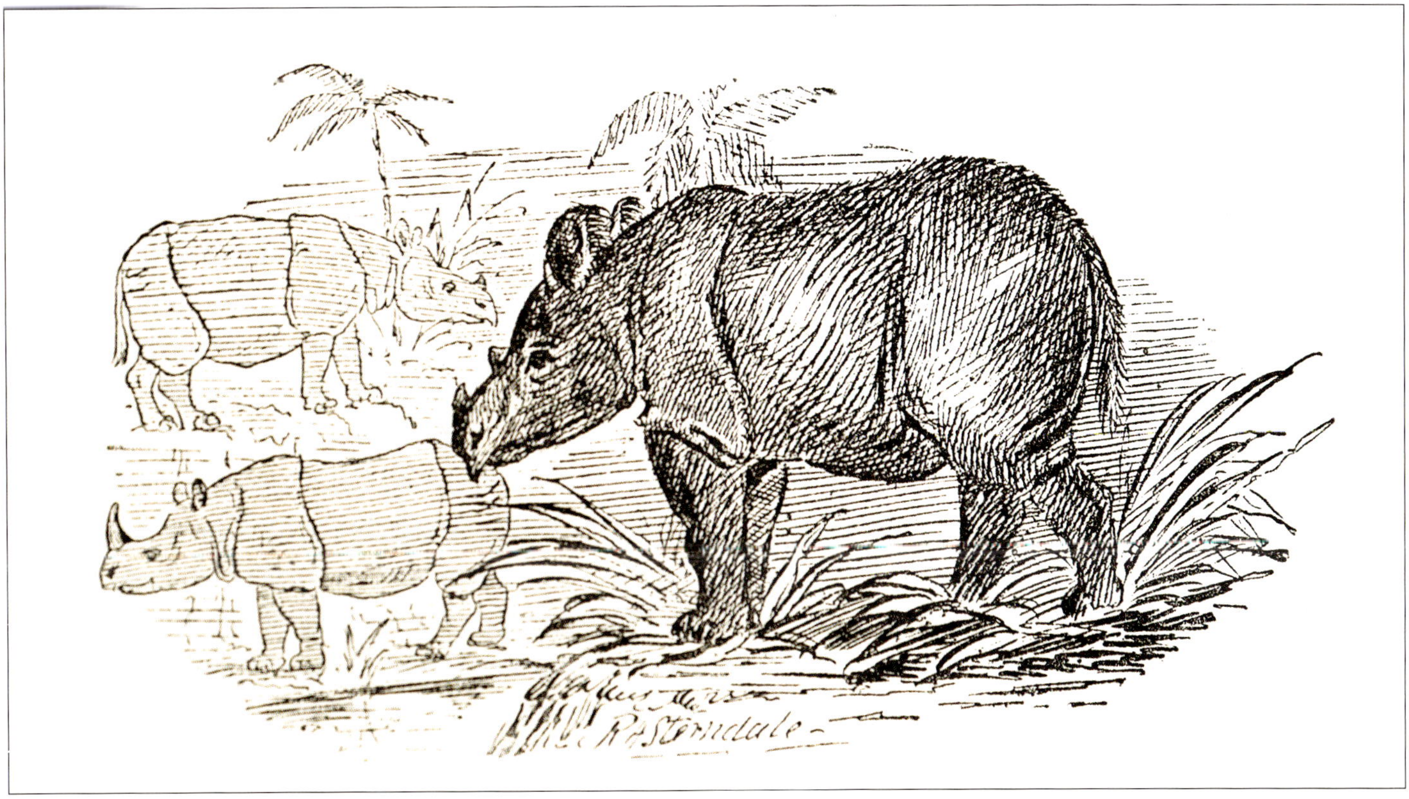

▪ A Sumatran rhinoceros with a Javan (earlier also known as Burmese) rhinoceros in the upper background and a greater one-horned (earlier also known as Indian) rhinoceros in the lower background.
From Sterndale, 1884.

In Sanskrit the rhino is called *khadga* which incidentally is also the word for a sword or scimitar, used in the *Mahabharata* and elsewhere. According to the *Maitrayani Samhita*, *khadga dhenu* is a female rhinoceros or a small sword.[12] Recently, in an attempt to follow Carl Linnaeus (1707–78) the founder of modern taxonomy, Sanskrit nomenclature has been coined to identify the three living Asian rhinoceros species. The greater one-horned rhinoceros has been named *ek-shring gandak* or *maha ek-shring gandak*. The Javan rhinoceros has been named *laghu ek-shring gandak* or *kamrup gandak* while the Sumatran rhinoceros is the *dvipi shring gandak* or *jambudvip dvi-shring gandak*.[13] While it is quite possible that the word *gandak* may have been used for all three rhinoceroses in ancient times, the latter nomenclature seeks to identify each of the three species separately. However, the Sanskrit names, ancient and modern, have been forgotten in popular parlance and they remain only on the dusty shelves of libraries.

The world over, all the five extant species of this animal are known by the word rhinoceros. The word is of Greek origin: *rhinos* means nose and *keratos* or *cero* is horn. Here the reference is to the horn being "composed of keratin, a substance derived from the skin, like fingernails".[14] The Javan rhinoceros is called *Rhinoceros sondaicus,* thus identifying the species as inhabiting the Malaysian islands known as Sunda islands and the Sunda straits which lie between Java and Sumatra. The Sumatran rhinoceros on the other hand was called *Didermocerus sumatrensis* and is now known as *Dicerorhinus sumatrensis*. The old genus name is derived from the Greek words *di* meaning two, *derma* meaning skin and *cero*, horn (there are two horns). The Latin suffix *ensis* means belonging to; thus the species name is self-explanatory.

The greater one-horned rhinoceros is known by its scientific name *Rhinoceros unicornis* given to it by Carl Linnaeus himself in 1758. In this case the species name *unicornis* is of Latin origin: *unus* meaning one; (genitive) *cornus*, the horn.[15] Thus a

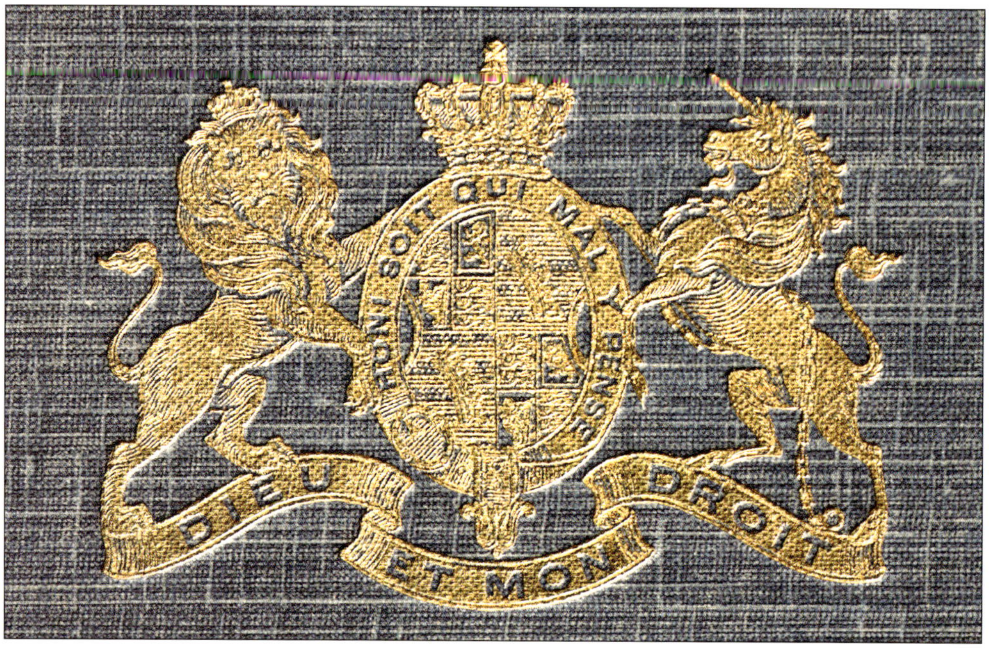

■ Royal coat of arms of the kingdom of England, in which the mythical unicorn is depicted as equine, as is common in European heraldry. The coat of arms became the symbol of the British Indian Empire.
From Curzon, 1925.

free English translation of the scientific name would be "single nose horned animal"! The word "unicorn" needs to be examined at some length to understand how it got fixed to our *gainda*. The *Oxford English Dictionary* states the following: "The unicorn has at various times been identified or confused with the rhinoceros, with various species of antelope, or with other animals having a horn (or horns) or horn-like projection from the head. According to Pliny [the Elder, 23–79 CE], it had the body of a horse, the head of a deer, the feet of an elephant, and the tail of a lion with one black horn projecting two cubits [18–22 inches x 2 = 36–44 inches, i.e. 91–112 cm] from the middle of the forehead."[16] Possibly, Pliny may have seen a rhinoceros brought from Asia (as he mentions one horn) for the Roman games or for display, though his description is not very accurate.

Aristotle (382–322 BCE) was much intrigued by the unicorn. He wrote in his classic work on Natural History: "But, a few animals are known to be single-horned and single-hoofed as the Indian ass; and one to whit the oryx is single-horned and cloven-footed."[17] Aristotle's description of the unicorn has been drawn from the accounts of classical authors who wrote on India and talked of a single-horned ass. There is no equid with a horn, and the animal described as a single-horned ass was probably the rhinoceros as interpreted by travellers to India of that time. The oryx would appear to have a single horn only if it was seen in profile and was recorded in a picture as such. However, the non-existent and mythical horned equid found pride of place in European heraldry as the unicorn, while more appropriately Linnaeus gave the name "unicornis" to the *gainda*. The *gainda*'s mythical counterpart would arrive in India to rule over it as the sinister supporter in the royal coat of arms where the dexter supporter was the ubiquitous lion which became the symbol of the British Indian empire. In a popular ditty though, the unicorn fought with the lion for the crown which surmounts the coat of arms!

Telling the Story of the Unicorn

The *gainda* is an animal which to a general observer looks like a creature left over from a long bygone era. The male animal is somewhat bigger than the female. Its body length is between 335 and 346 cm, its tail being 66 cm long; with a shoulder height of 175 to 200 cm, it weighs around 2,000 kg. To quote Eric Dinerstein: "This species has two distinct skin folds that drape over the sides of the front and hind legs, deep scars on anal folds, extra skin known as epidermal knobs, cuts on the ears, and missing tail tips. Body colour is grey and mosaic-like pattern on the skin is quite noticeable on the rump."[18]

The most important point for our story is its horn which is without a bony core and "contains tiny keratin tubes dispersed in a keratinous matrix",[19] growing over its nose. *Rowland Ward's Records of Big Game* has catalogued 23 horns of more than 11 inches (28 cm), the longest recorded being 24 inches (61 cm).[20] Having written all this, one must accept the fact that the animal defies description and it has to be seen to fully appreciate its prehistoric appearance.

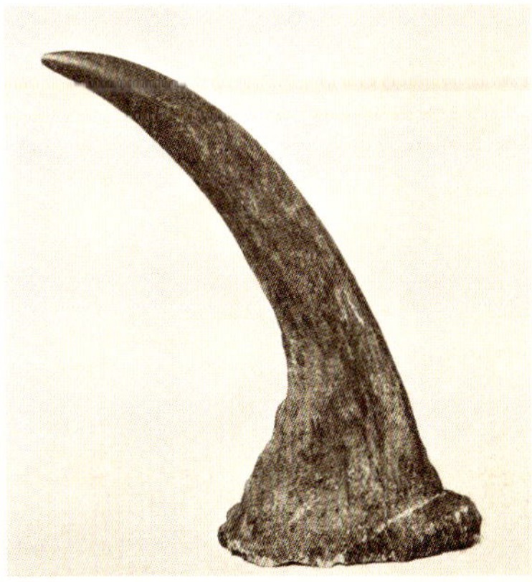

Record Horn of Great Indian Rhinoceros.

■ A record horn of the greater one-horned rhinoceros in the Natural History Museum, London. Its length is not mentioned, but the longest they have recorded is 24 inches (61 cm). *From Dollman and Burlace, 1935.*

The Indian species of the rhinoceros is primarily a grazer. It inhabits, today, the protected areas along the large riverine tracts along the Himalayan mountain range. It lives amidst the tallest grass in the world, popularly called elephant grass because of its height. Today, the most famous abode of the greater one-horned rhinoceros is the Kaziranga National Park along the Brahmaputra river system in Assam. The environs of oxbow lakes and other waterbodies are its favourite feeding and wallowing grounds.

During the monsoon, though, when Kaziranga gets waterlogged, the rhinos move into the adjoining hills, returning to Kaziranga as the waters recede. The males are territorial, marking their area with urine and dung, and charging aggressively at male intruders. The females carry their young for 16 months and their calves remain with them for three to four years.

While history writing on the environment came into its own in the second half of the 20th century, few works have focused on the history of human interaction with a particular species. From Europe, the masterly work of Michel Pastoureau comes to mind. He chronicled the European brown bear's unfortunate brush with humanity leading to its near extinction there. From Japan, Brett L. Walker wrote the story of the extinction of the wolf; and Andrew C. Isenberg chronicled the story of the near extinction and resurrection of the iconic bison by the white man in America.[21] In India we are not far behind. The ubiquitous goose's presence in art and literature has been chronicled by Jean Philippe Vogel and the peacock has found a biographer in P. Thankappan Nair.[22] Coming to animals, the cheetah, the lion and the elephant have been chronicled extensively.[23] Several tomes have been written on the charismatic tiger, yet it awaits a serious biographer. So does India's unicorn, and our story proposes to fill this gap.

To flesh out information on human-animal interaction through the ages is a task which involves meandering through a variety of sources from different time periods. Some animals being iconic, like the lion and the elephant in Asia and Africa or the brown bear in Europe, are found to be fairly common in various sources. The rhinoceros on the other hand, though a very noticeable large mammal, is not quite in the same league.

The rhinoceros appears in cave and rock art of the Indian subcontinent before the written word. Through examination of such depictions, its presence in geological times since the Pleistocene age has been established. Shibani Bose presents the prehistoric evidence in chapter 2, while in the next chapter she provides a portrait of the animal through Harappan seals, literary sources, medical texts of ancient India,

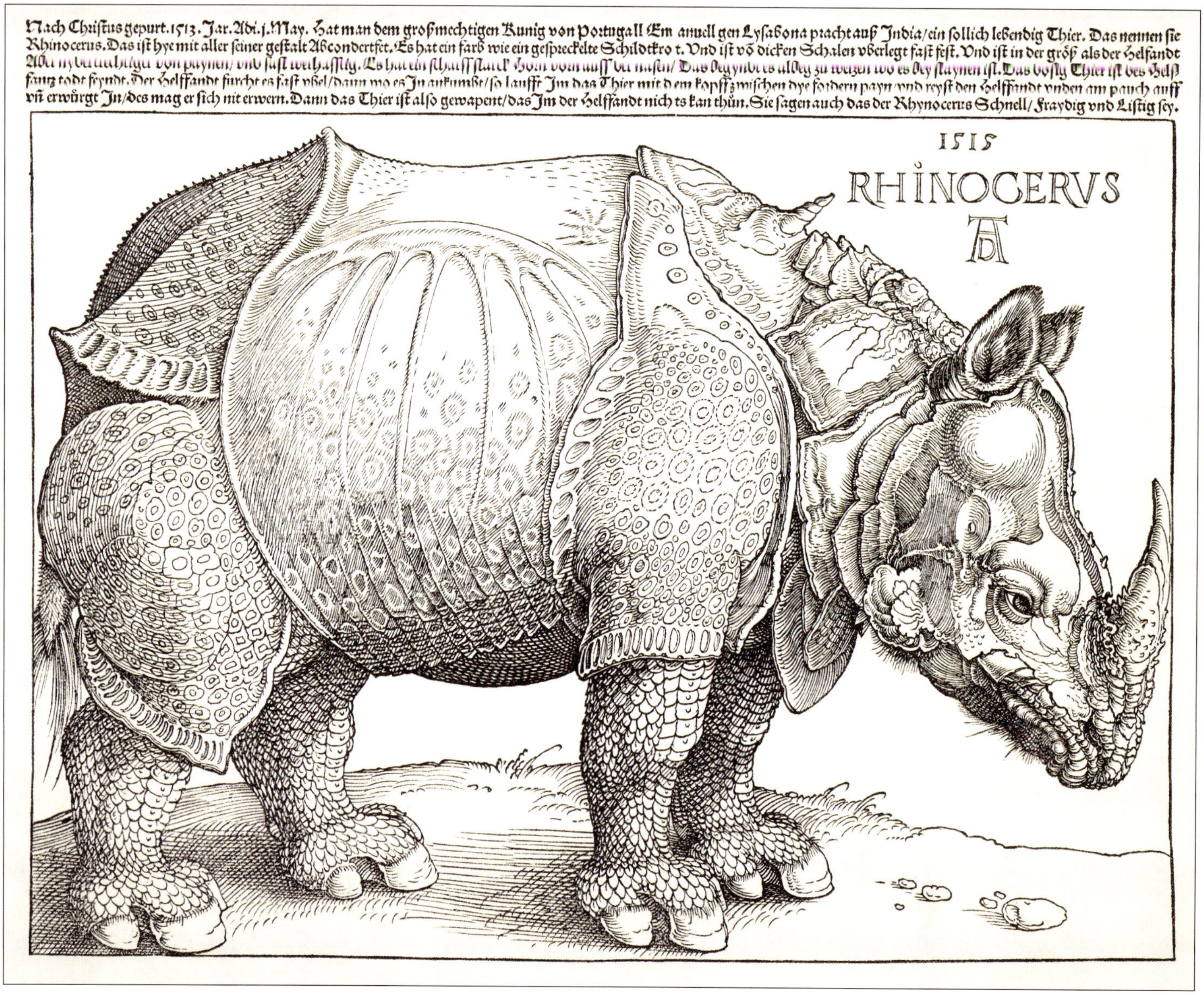

■ A fanciful rhinoceros: Albrecht Dürer (1471–1528), the German artist, drew this animal in 1515 from descriptions he had heard. He endowed the animal with an additional horn between its shoulders! *© The Trustees of the British Museum.*

sculptures at Sanchi and gloriously in Gupta coinage. These sources amply testify to the range of the rhinoceros in ancient India.

Between this early period and the coming of the Mughals in the 16th century, there is a major gap in the narrative which needs to be addressed here briefly. Apart from literary and other related sources of the time, the genre of travelogues and history writing by foreign visitors to India assumes importance in our story.

Foremost among the early travellers who provide information on the rhinoceros is Abu Rahim Al-Biruni. Originally from Khiva in modern-day Uzbekistan, he spent several years in Punjab around 1030, having accompanied the raiding armies of Mahmud of Ghazni to India. Al-Biruni studied Sanskrit and translated some works into Arabic, besides writing a detailed account of India and its people. He has left us a description of the rhinoceros as well: "The *ganda* exists in large numbers in India, more particularly about the Ganges. It is of the build of a buffalo, has a black scaly skin and dewlaps hanging down under the chin. It has three yellow hooves on each foot, the bigger one forward, the others on both sides. The tail is not long; the eyes lie low, further down the cheek than is the case with all other animals. On the

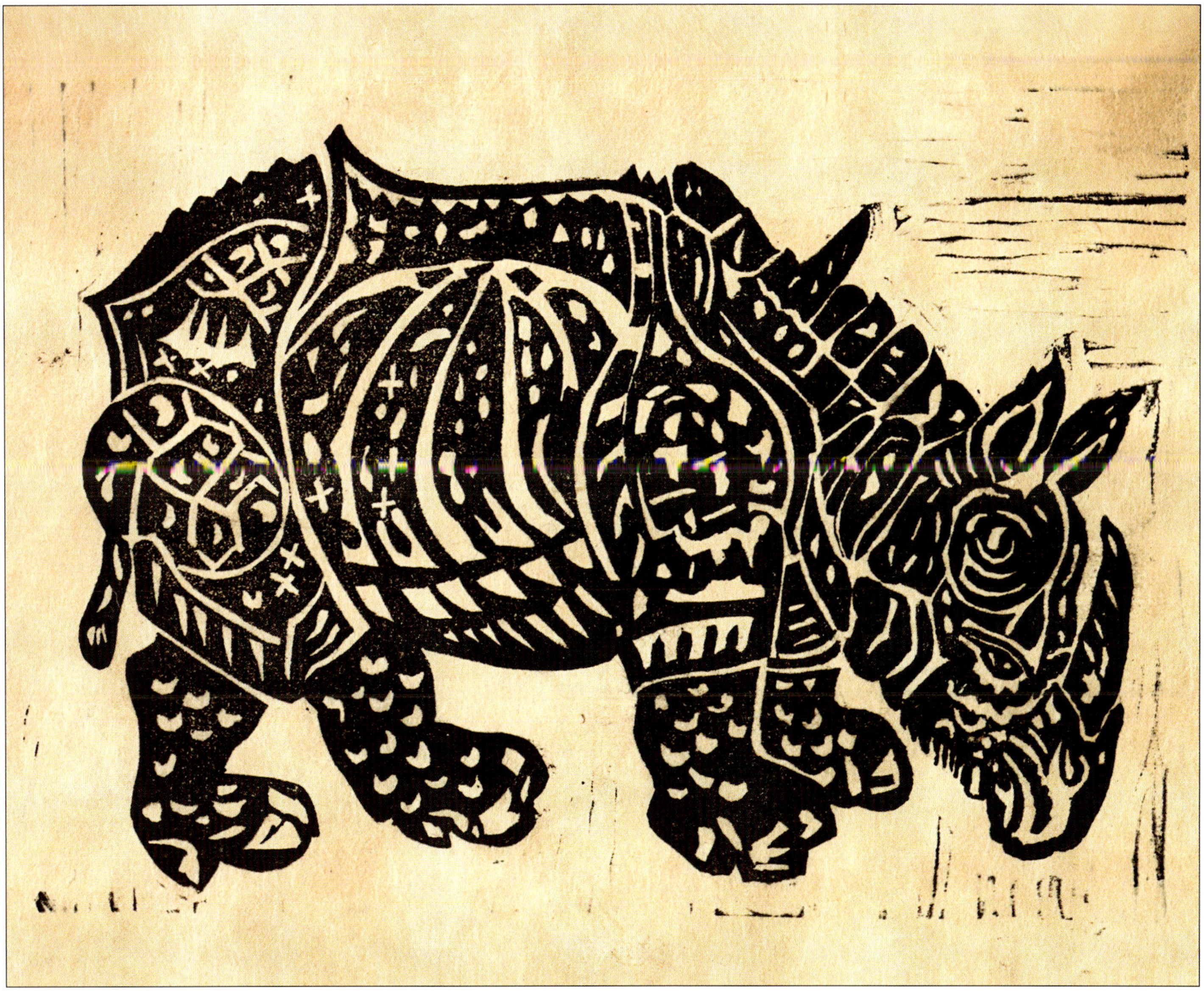

top of the nose there is a single horn which is bent upwards. The Brahmins have the privilege of eating the flesh of the *ganda*. I have myself witnessed how an elephant coming across a young *ganda* was attacked by it. The *ganda* wounded with its horn a forefoot of the elephant, and threw it down on its face."[24]

Three centuries later we have Muhammad Ibn-Battuta from Tangier who travelled to India and lived in the country from 1332 onwards. He saw India first-hand over two decades until 1353. He too was fascinated by the rhinoceros and has left a detailed description of the creature: "After crossing the river Sind [in the land] called Ban Jab [Punjab] we entered a forest of reeds ... when we were confronted by a rhinoceros. In appearance it is a black animal with a huge body and a disproportionately large head. For this reason it has become the subject of a proverb, as the saying goes *Al-karkaddan ras bila badan* (rhinoceros, head and no torso). It is smaller than an elephant but its head is many times larger than an elephant's. It has a single horn between its eyes, about three cubits in length [54–66 inches, i.e. 137–167 cm] and about a span in breadth. When it came out against us one of the horsemen got in its way; it struck the horse ... pierced his thigh and knocked him

▪ An illustration of a rhinoceros by Nandalal Bose (1882–1966) for *Sahaj Path*, a Bengali language learning book by Rabindranath Tagore, circa 1950. The depiction has interesting similarities with Dürer's rhinoceros, notably the shoulder-horn.

Private collection. Photograph © Christie's Images/Bridgeman Images.

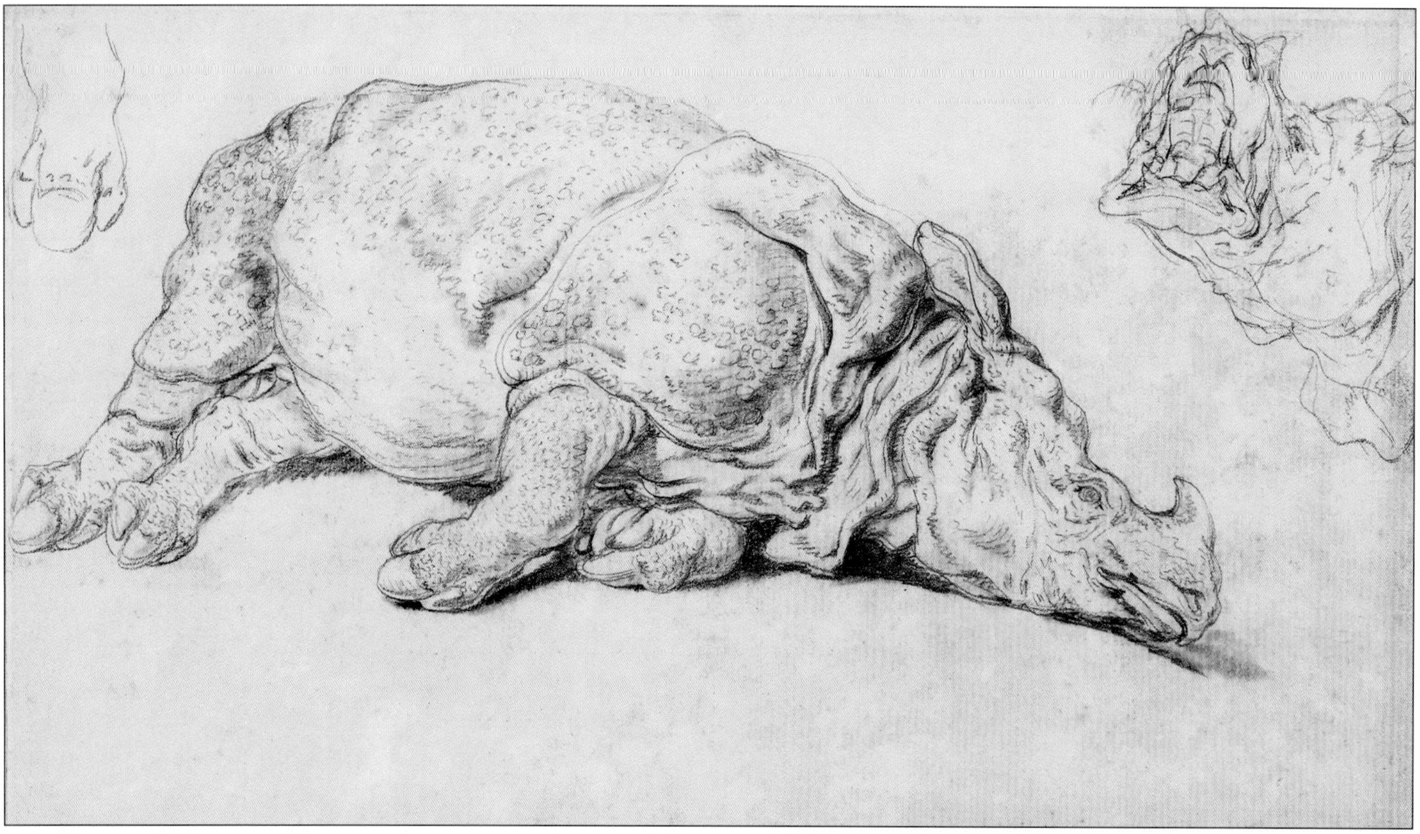

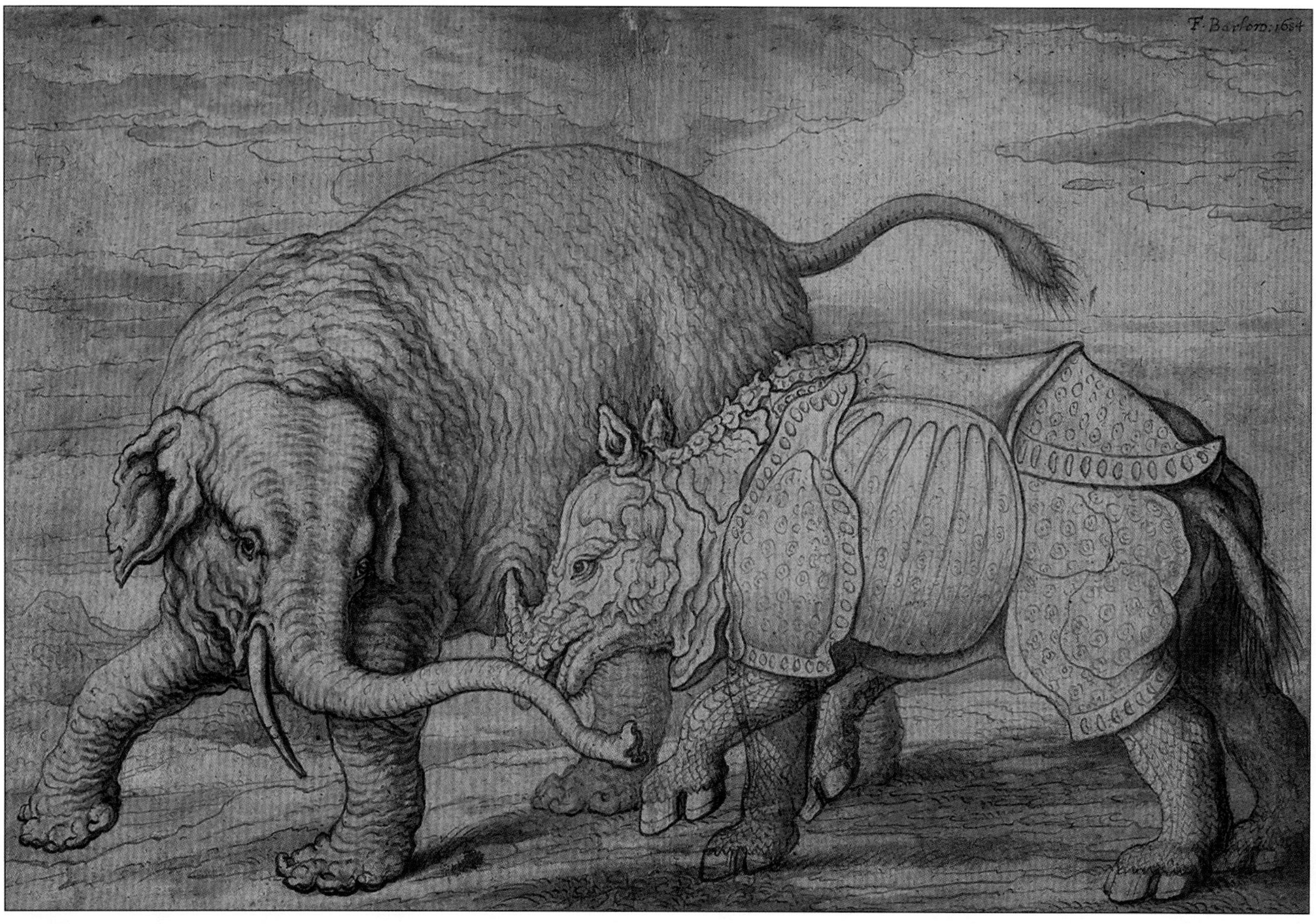
F·Barlow:1684

down, then went back into the thicket and we could not get at it. I saw a rhinoceros a second time on this road after the hour of afternoon prayer. It was feeding on plants but when we approached it ran away. I saw a rhinoceros yet another time when in the company of the King of India [Ghiyas ud-Din Muhammad Shah I Tughluq] we had entered a jungle of reeds. The Sultan was mounted on an elephant and we too were mounted on elephants along with him. The foot soldiers and horsemen went in and beat it up, killed it and conveyed its head to the camp (*mahalla*)."[25]

Al-Biruni's description is remarkably accurate, but Ibn-Battuta has allowed his fancy to take hold of his observations when he says that the head of a rhinoceros is bigger than that of an elephant. He does not mention the locality where the king hunted, but it could be the environs of present-day Delhi. We have yet another interesting notice of the rhinoceros, this one dating to 1387–88 from Prince Muhammad Khan, son and successor of Sultan Firuz Shah Tughluq. He marched towards the Sirmaur hills "and there spent two months, hunting rhinoceros and elk [deer]".[26] The locality is the Himalayan foothills in the southeast of present-day Himachal Pradesh.

With the establishment of the Mughals in Hindustan, a whole new window opens up to give us a glimpse of the life and times of the rhinoceros. Biographies of the great Mughal emperors and the autobiography of Jahangir, detailed chronicles of the administration of the empire, the history of the house of Timur, translations of the great epics *Mahabharata* and *Ramayana* from Sanskrit into Persian, all provide a plethora of sources to examine. In addition to literary materials, the profusion of paintings suddenly gives us an ocular record hitherto unparalleled.

Asok Kumar Das examines these sources in Chapter 4, to record the rhinoceros's brush with the empire, from the arrival of Babur in 1526 through Akbar's and Jahangir's shikar to Aurangzeb during whose reign an album of paintings of animals was made which includes a remarkable picture of the rhino. In the subsequent chapter Das takes us on a tour of Jain iconography, paintings of the *Bhagavata Purana*, depictions of the hunting expeditions of Bundi kings and of the rhinoceros in the Mewar school of painting. Though the animal did not exist in Rajasthan at the time, the Bundi kings may well have hunted them elsewhere, and the Mewar paintings are tantalizing. Das presents a variety of sources, from Jean-Baptist Joseph Gentil's album which records the Nawab of Avadh's rhinoceros hunt near Bahraich (present-day Uttar Pradesh) in the late 18th century, Thomas Daniell's exquisite rendering of the unicorn from the foothills of the Himalaya, the rhinoceros at the Peshwa's court at Pune, to Indo-Portuguese rhino-horn artefacts and a temple relief in Murshidabad, West Bengal. These sources go further to establish the animal's existence in fair numbers in its range up to the beginning of the 18th century, apart from creating a firm niche for its presence in the cultural world of the period.

Before we come to the British period, a short diversion to make note of the rhinoceros's arrival in Europe is worth taking here. We have seen earlier that there was a huge misconception of the unicorn in the ancient world of Greece and Rome. It appears that no accurate descriptions of the rhinoceros were available in Europe even in the 16th century, as is testified by the rhinoceros drawn by Albrecht Dürer in 1515. He endowed the animal with a small horn protruding from between its shoulder blades, apart from a very stylized body.

In 1581 a living specimen brought from India by the Portuguese was presented to King Philip II of Spain and came to be known as the "Madrid" rhinoceros. In

▪ The "Dutch Rhino": this is an accurate rendering of Clara at rest by the German painter Johann Elias Ridinger (1698–1767), dated 1748.
Courtesy The Samuel Courtauld Trust, The Courtauld Gallery, London.

▪ "Rhino-Elephant Fight": an imagined rendering of the event by English artist Francis Barlow (1626–1704), dated 1684.
Courtesy The Samuel Courtauld Trust, The Courtauld Gallery, London.

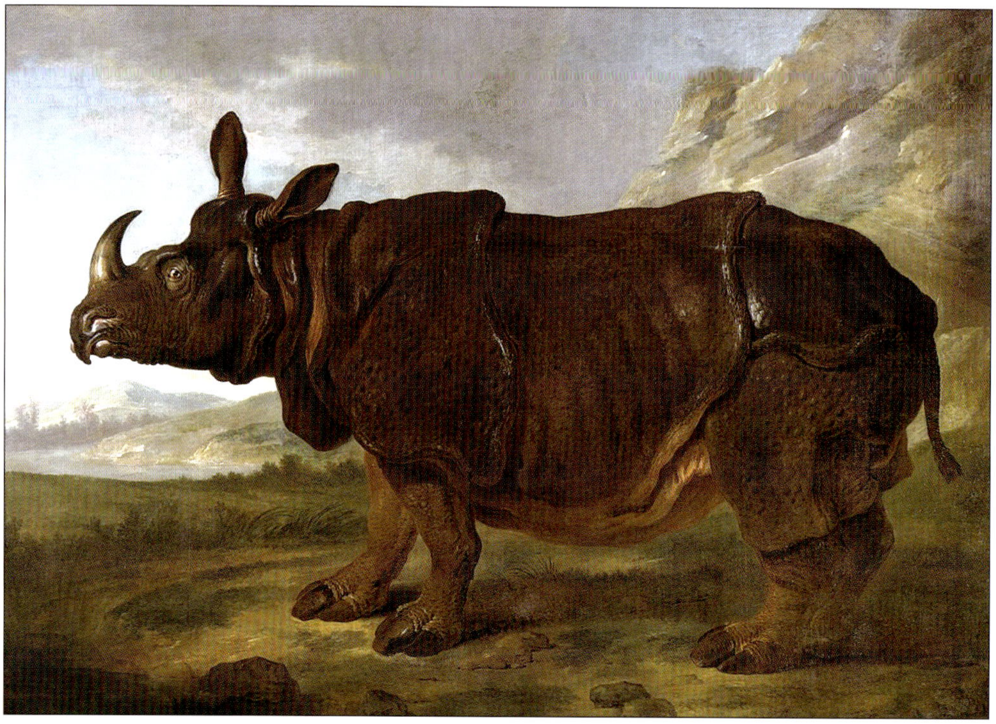

■ "Clara le Rhinoceros": Louis XV's favourite painter, Jean-Baptiste Oudry (1686–1755), painted the famous female rhinoceros during her stay in 1749.
Source: wikiart.org.

1684 an engraving by Francis Barlow claimed to be "a true representation" of a fight between a rhinoceros and elephant, the rhino having been "lately brought over from the East Indies to London". However, the rhinoceros here is suspiciously similar to Dürer's, though without the horn on its back. Dürer's depiction also seems to be the inspiration for the 20th-century Bengal school artist Nandalal Bose in his stylized linocut illustration of circa 1950.

In 1741 a Dutch sea captain, Douwemont Van der Meer, bought young Clara, a female rhinoceros, from J.A. Sichterman, a Director of the Dutch East India Company in Calcutta.[27] The travails of this poor creature need not be detailed here, except to say that she was transported from Calcutta to Rotterdam via the Cape of Good Hope, and over the next few years toured various royal courts in Europe, from Paris, London and Berlin to Krakow, Prague, Rome and Naples. The "Dutch" rhinoceros was a curiosity and earned the sea captain a comfortable living until her death in London in 1758. Of interest to us here is the accurate rendering of the species in Europe, finally: Clara was immortalized by the German painter Johann Elias Ridinger in 1748, and by King Louis XV of France's favourite artist, Jean Baptist Oudry in 1749, among others.

As early as 1684, a Captain Henry Udall brought a rhino to London on his ship. It was from "the Court of the King of Gulkindall", i.e. Golconda. A second one reached London in 1739. One rhino arrived in France in 1770 and another landed in London in 1790 known as the "Stubbs" rhinoceros since it was painted that same year by the famous artist George Stubbs, and yet another was brought to London in 1799.[28] Many more were to follow.

The fate of these animals has been meticulously recorded in detail by T.H. Clarke. They were curiosities in 18th-century Europe where the memory of the Ancient Roman games that had included African and possibly Indian rhinoceroses had long faded. These animals were objects of wonder, kept in the menageries of kings and potentates, and made a remarkable impression in European art, tapestries, porcelain

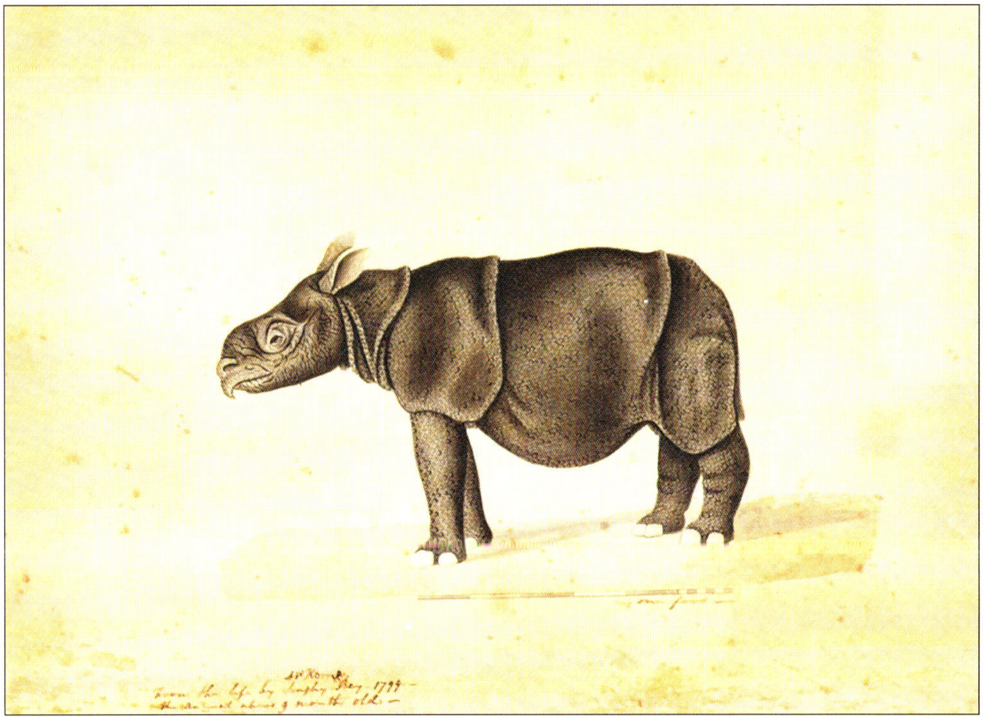

■ "Young Indian Rhinoceros" by Robert Home, watercolour on paper, 1799. The inscription by John Fleming mentions that the animal is about nine months old.

By kind permission of the Trustees of the Victoria Memorial Hall, Kolkata, Fleming Collection (R 2609/53).

and clocks. Some of them travelled from one city to another being exhibited and earning handsome amounts for their owners.

However, important as these sojourns of India's unicorns were to European art and culture, they had little to contribute to the animal's interaction with man in its own homeland. These unfortunate animals had to face captivity, voyage through choppy oceans and life in captivity in an alien environment, but what is relevant here is the rather accurate renderings of them by European artists as mentioned above.

Coming back to our story of the unicorn in India, Chapter 6 on the British period and beyond traces the travails of the animal which had all but disappeared from its former range except for some areas in the easternmost area of the subcontinent, where it precariously hung on to existence near extinction. A British surgeon, John Fleming, who was President of the Bengal Medical Service in 1800, assembled albums of over 200 natural history studies of local fauna by European and Indian artists.[29] Notable for our context is a study of a young Indian rhinoceros by Robert Home (c. 1799). However, as in the Mughal period, most of the images from this time relate to hunting. We give an account of the relentless shikar in colonial times, and later conservation efforts by the British Indian government and the government of independent India which led to the successful stabilization of its population and the increase in its numbers from less than 200 around 1900 to a total population of more than 3,500 between Nepal and India in the 21st century.

The final chapter examines the continuing efforts at conservation up to the present time, including translocation of the animal within its former range. It also examines some aspects of the most serious threats to its existence caused largely by the demand for its horn which is believed to have medicinal value. India's unicorn still lives on the brink, notwithstanding its successful conservation that appears to have made its survival secure for the moment.

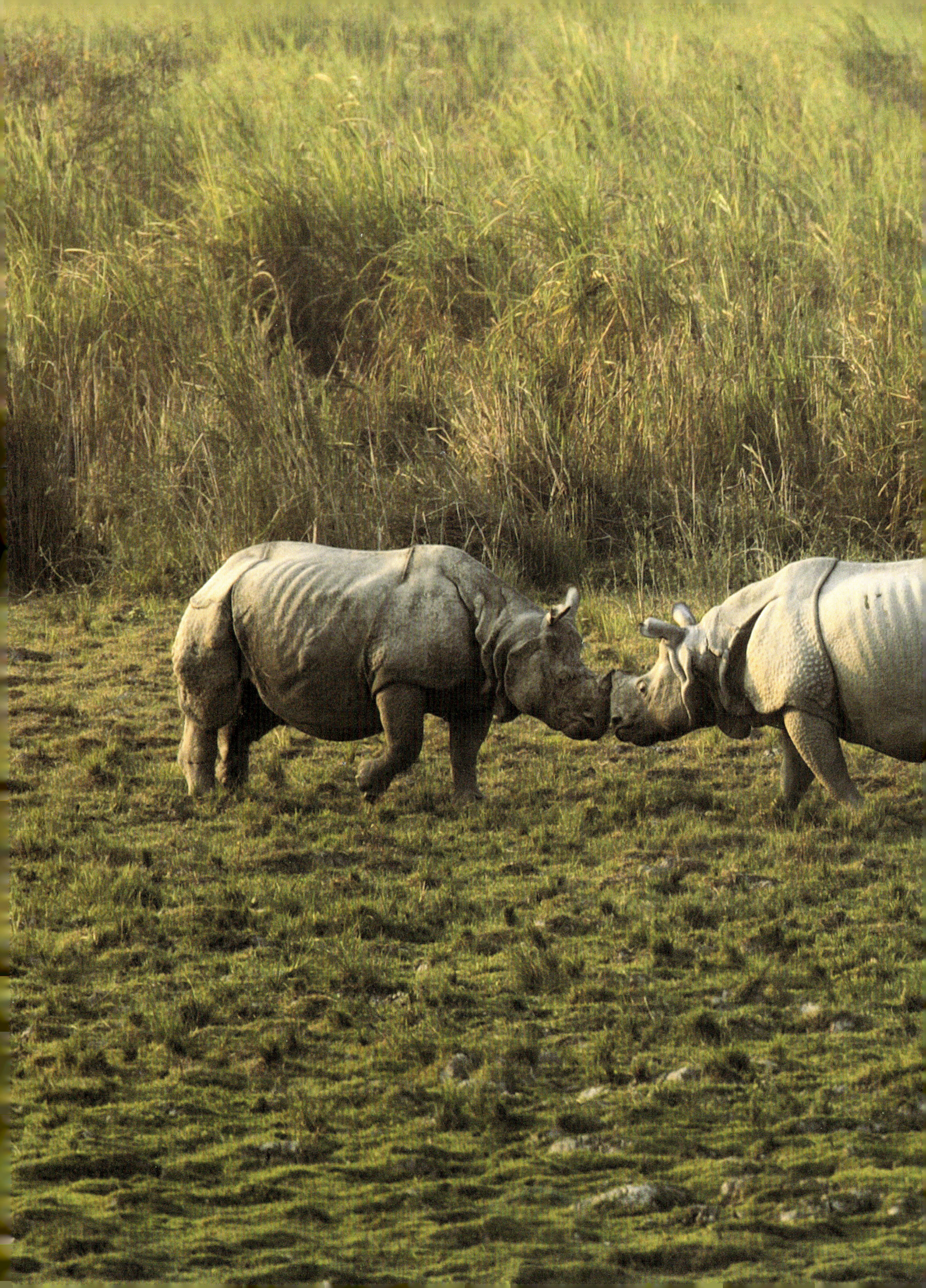

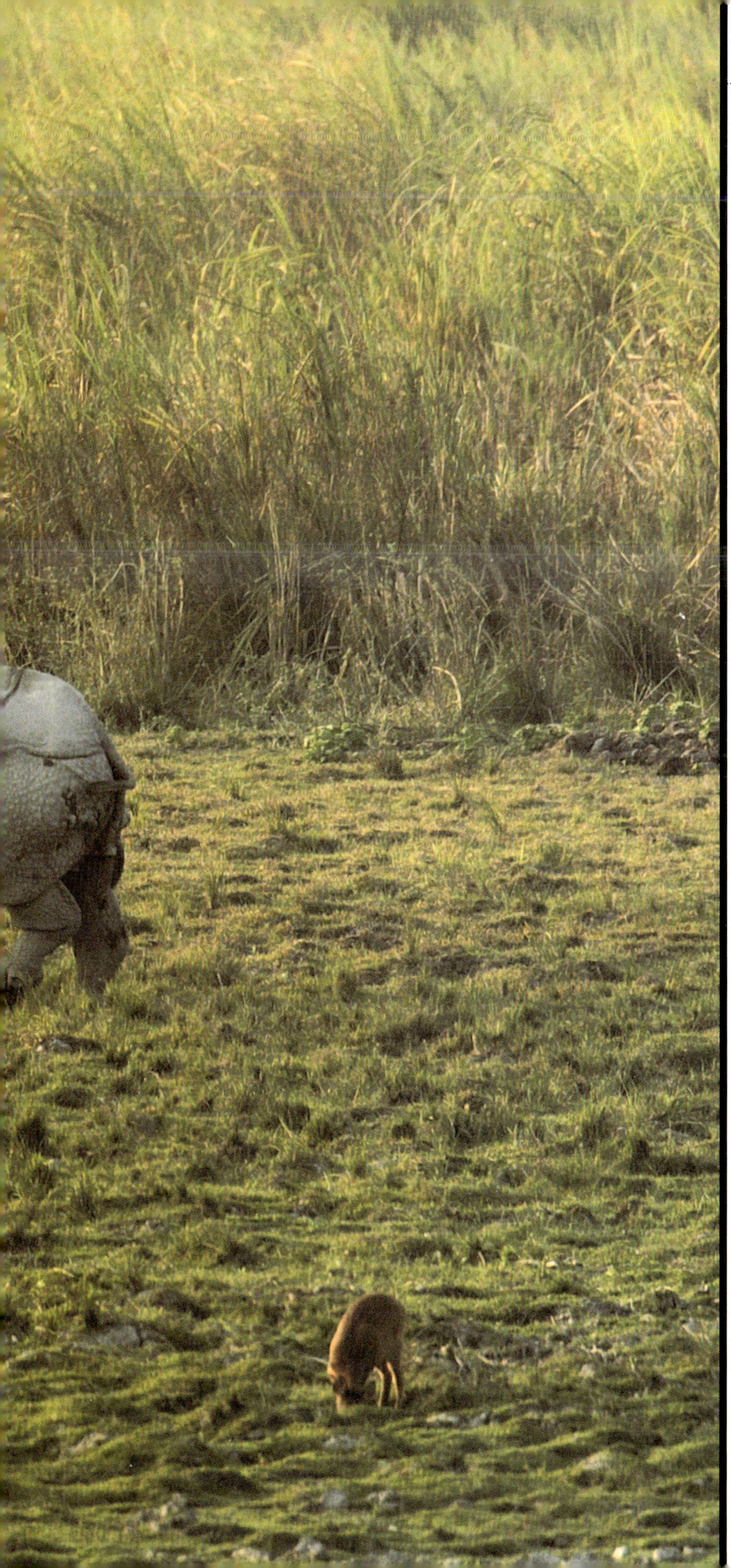

■ Being territorial in nature, a male rhino will protect its domain from intruders. Note the tall elephant grass of Kaziranga.
Photograph: Kakubhai Kothari.

CHAPTER 2

Before the Written Word

Shibani Bose

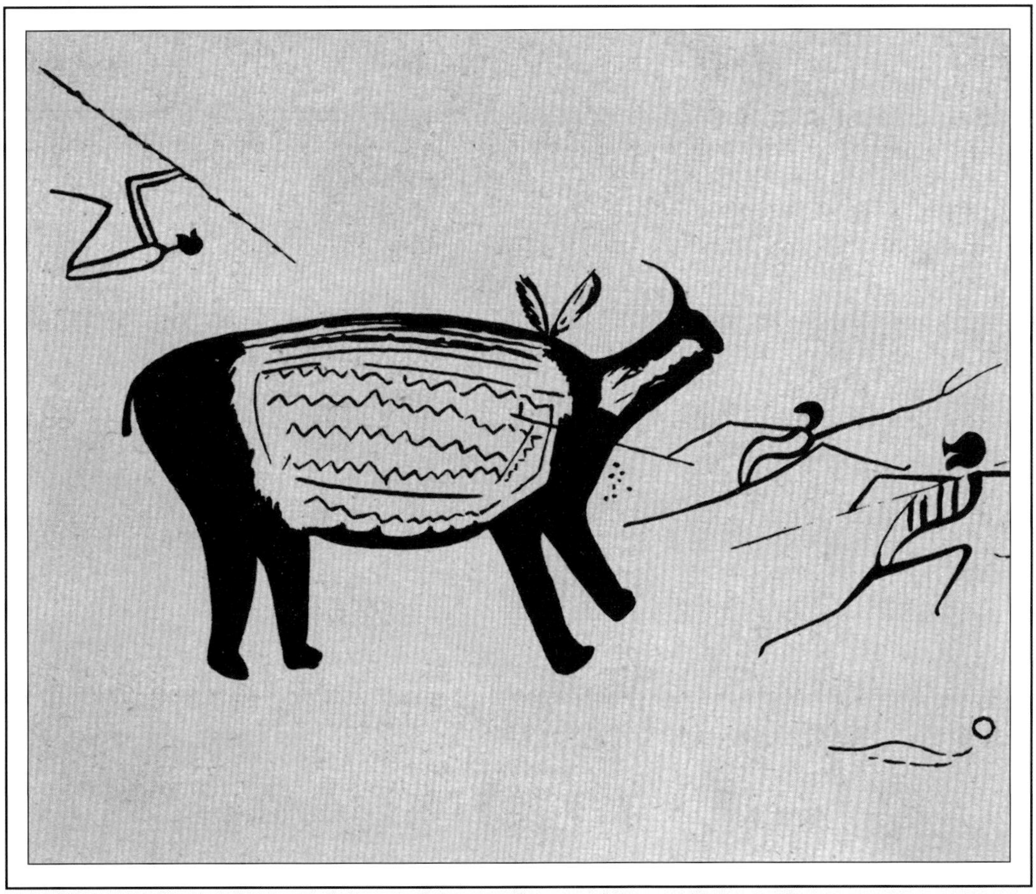

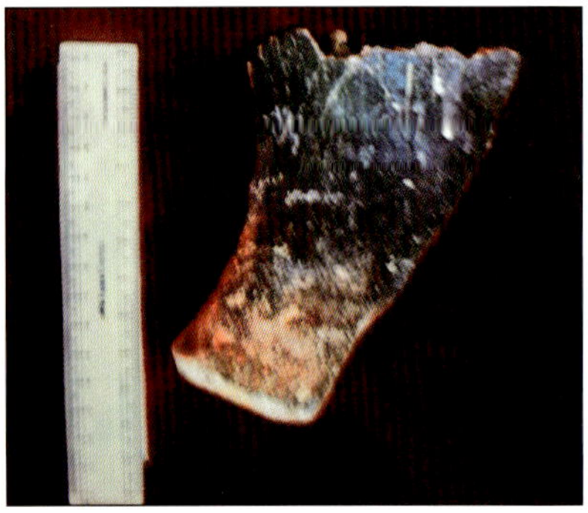

■ Fossilized rhino horn core from the late Pleistocene period, found in the central Narmada valley.
Photograph courtesy G.L. Badam.

LUMBERING ITS WAY ACROSS the tapestry of time, the rhinoceros has significant antiquity in India. But how do we go about reconstructing this story? The animal, like most others, can be looked at through multiple prisms, ranging from faunal remains and visual depictions retrieved from the archaeological record to the formidable corpus of ancient Indian literature. Carefully sifting through an array of sources that have a bearing on our one-horned protagonist, we may reconstruct the story of the animal's journey across millennia before it disappeared from most regions of the Indian subcontinent.

Faunal Remains

The earliest traces are in the form of fossil remains retrieved from Indian geological formations. Fossil fragments tell tales of numerous extinct species that roamed the subcontinent. What we know is that amidst a plethora of such forms, the recognized species of the genus *Rhinoceros* are *Rhinoceros sivalensis*, *Rhinoceros palaeindicus* and *Rhinoceros platyrhinus* from the Upper Siwaliks and *Rhinoceros perimensis* from the Middle Siwaliks. Similarly, we know of *Rhinoceros deccanensis* from the Krishna valley near Gokak on the Ghataprabha river in Belgaum district, Karnataka, and *Rhinoceros karnuliensis* from the Kurnool caves, Andhra Pradesh, in late Pleistocene contexts.[1]

Though the genus itself can be traced back to the Pliocene of northern India, most known fossils of *Rhinoceros unicornis*, the species we are looking for, seem to be from the Pleistocene.[2] Fossil and semi-fossil remains of the species retrieved from the Narmada valley in central India along with those from peninsular India are crucial pointers to the past distribution of the species whose geographical retreat to its present-day haven is a fairly recent phenomenon.[3] For instance, in a late Pleistocene context dated to 40,000 BP, three horn cores of the unicorn are reported from the gravel deposits of the Sher river tributary in the central Narmada valley. The significance of these retrievals lies in the fact that this may be the only find of a fossilized horn core not just in India, but possibly also in South Asia.[4] Similarly, the partial skull of the animal in a late Pleistocene to Holocene context at Sathankulam in Tirunelveli district of Tamil Nadu, is the first record of fossil *Rhinoceros* from the southern tip of India. The mega mammals, which have totally disappeared from south India, possibly inhabited this area when the low hills were forested with swamps in short canyons surrounded by grass.[5]

Perhaps one of the first accounts to report evidence of the presence of the rhinoceros in conjunction with human activity is that by John Cockburn.[6] A British officer in the Opium Department, Cockburn was on a hunt in July 1881, in the ravines of the Ken river, two miles (3.2 km) south of the town of Banda in Uttar Pradesh, when he chanced upon the fossil remains of a rhinoceros, assigned to the species *Rhinoceros indicus*. The remains he reported included a large quantity of fragments of teeth along with some pieces of longitudinally split shafts of long bones. Within four feet of the rhino bones, Cockburn also found knives made of chert and shell on the surface of the soil.

What makes Cockburn's finds even more significant is that he simultaneously drew attention to a rock painting in the Ghormangar rock shelter in Mirzapur district, Uttar Pradesh, representing a rhino hunt. While he seemed to consider this painting to be of a recent date—not more than 300 years old—the fact that the painting depicts the use of harpoon-like weapons probably made of wood and stone, and that he found stone implements near the rhino remains, suggests that this was likely to be an ancient site whose antiquity eluded Cockburn.

In the context of the subcontinent as a whole, the cohabitation of ecosystems by humans and the rhinoceros can, however, be pushed back much further. Such a presence, for instance, has been unearthed in the terminal Pleistocene in the middle Son valley. While stone artefacts strongly testify to the presence of humans in the region during the terminal Pleistocene, R.J. Blumenschine and U.C. Chattopadhyaya reported the presence of the rhinoceros in the terminal Pleistocene fossil faunal assemblage of the middle Son valley.[7] The inference that the species was *Rhinoceros unicornis* was based on the presence of the animal in this region of India during the late Pleistocene.[8]

Here, then, were the beginnings of an interface which was to mature and manifest itself in the realms of subsistence as well as aesthetics. Since then, archaeozoological research has enabled us to document the presence of the unicorn in various cultural niches from hunter-gatherer societies to the first urban civilization of India and beyond. The details of the emerging narrative will not only be integral to recovering echoes of the animal's presence in areas where it is now extinct, but also for assessing the cultural dimensions of the interface the armoured giant had with humans in the ancient past.

Following its appearance in the fossil record, the unicorn steadily makes its way onto the canvas of early India in the form of tangible remains retrieved from prehistoric and protohistoric sites. Bones tell their own tales, and it is these that we look for when we use animal remains as one of the many prisms which can be employed to chart larger faunal histories. Not only do such records help us in mapping the past distribution of the species within the subcontinent, they also help us envision a landscape vastly different from the one we see today.

As far as the archaeological record is concerned, one proceeds with the consciousness that the evidence is scattered, and at times, disappointingly patchy. As more sites are excavated and their faunal assemblages examined, we hope to know much more about human interactions with the mega mammal. Till then, even on the basis of the evidence available, it is possible to discern traces of the animal's footprints in time, and to connect those vestiges to give it the history it deserves.

We begin our quest with the microlith-bearing sites in India, several of which are mesolithic. Microliths are stone tools that can measure anywhere from less than 1 cm to 5 cm, and are fashioned out of varieties of fine-grained stones. It is important to clarify at this point that though considered a hallmark of mesolithic cultures, the use of microliths is not confined to them. This implies that while every mesolithic site is likely to yield microliths, the latter as tool types do not necessarily suggest a mesolithic stage.

The first cluster of mesolithic sites that testify to the presence of the rhinoceros occurs in the Middle Ganga Plain. It is known that oxbow lakes and other water features are important for the animal in view of its need to wallow and feed on aquatic plants.[9] Hence, it is a matter of little surprise that sites in the region have yielded evidence for the presence of the animal in early contexts. Archaeology tells us that the Middle Ganga Plain was characterized by a large number of horseshoe or oxbow lakes, most of which have now been filled up by natural and human agencies and converted into farmlands.[10]

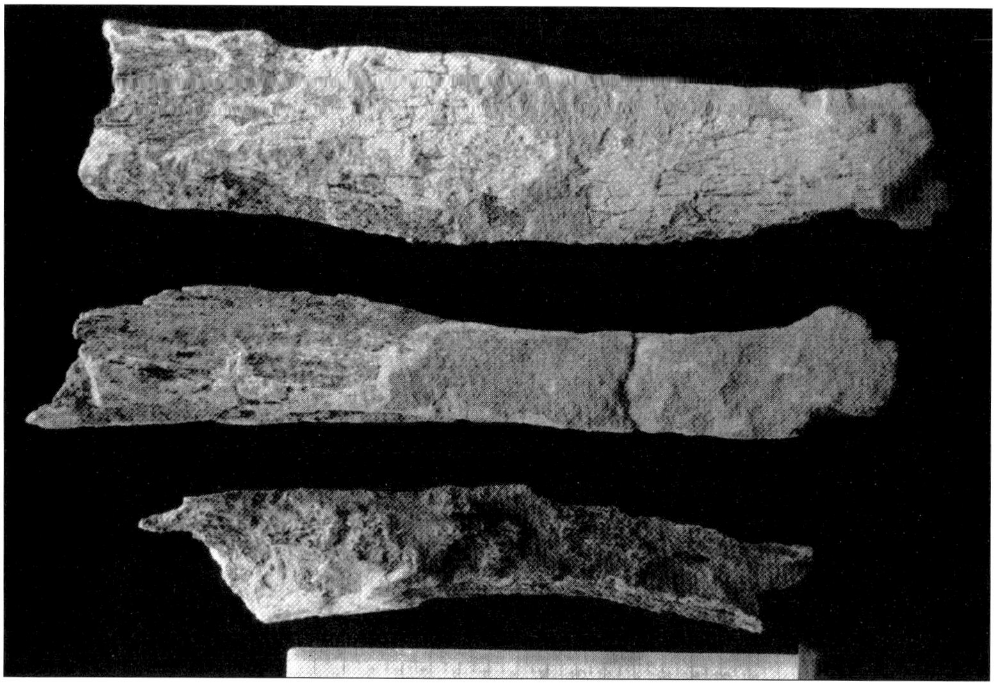

■ Three fragments of rhinoceros rib bones from Langhnaj, Gujarat.
From Sankalia, 1965, plate XII a.
© Deccan College, Pune.

Overall, palaeovegetational studies suggest that grassy forests were certainly more widespread in the past, and gave shelter to a number of herbivores such as different species of deer, rhinoceros and wild boar.[11] Pollen analysis together with faunal evidence testifying the presence of such species suggests a mosaic environment in the past—forest, grassland, and marshland.[12] The landscape that archaeology reconstructs for us, then, seems to have been one ideal for a moisture-loving species like the Indian rhinoceros which is rarely found more than 2 km away from water.

Perhaps the most ancient of sites in the Middle Ganga Plain is Damdama in Pratapgarh district of Uttar Pradesh. Its antiquity is tentatively placed in the first half of the 7th millennium BCE.[13] The site yielded bones of large mammals like the elephant, rhinoceros, gaur, wild buffalo and possibly wild cattle. However, the hunting of such large mammals for purposes of food was considered unlikely in view of the technological level of the mesolithic population.

Instead, it was conjectured that the carcasses or isolated bones were collected and utilized for making bone tools. It was also observed that the bones of large mammals were concentrated on the eastern part of the site, and that these bones were well preserved, without much charring or breakage. A majority of the bones revealed cut marks, leading to the inference that the bones possibly had been cut off from the original carcass and dumped at one place near the settlement for later use. These were, therefore, postulated to have been intentionally kept raw material for the preparation of bone tools and objects.[14]

In the vicinity of Damdama, we encounter the rhinoceros again at Sarai Nahar Rai and Mahadaha. The former, radiocarbon dated to circa 8395±110 BCE, revealed hearths and floors yielding burnt clay lumps along with charred as well as uncharred animal bones and microliths. These bones indicated the presence of stag, bison and rhinoceros.[15]

Unlike Damdama and Sarai Nahar Rai, the dating of Mahadaha is fraught with uncertainty.[16] Nevertheless, the site yielded definite evidence regarding the presence of the rhinoceros in the form of remains including teeth, a mandible and a

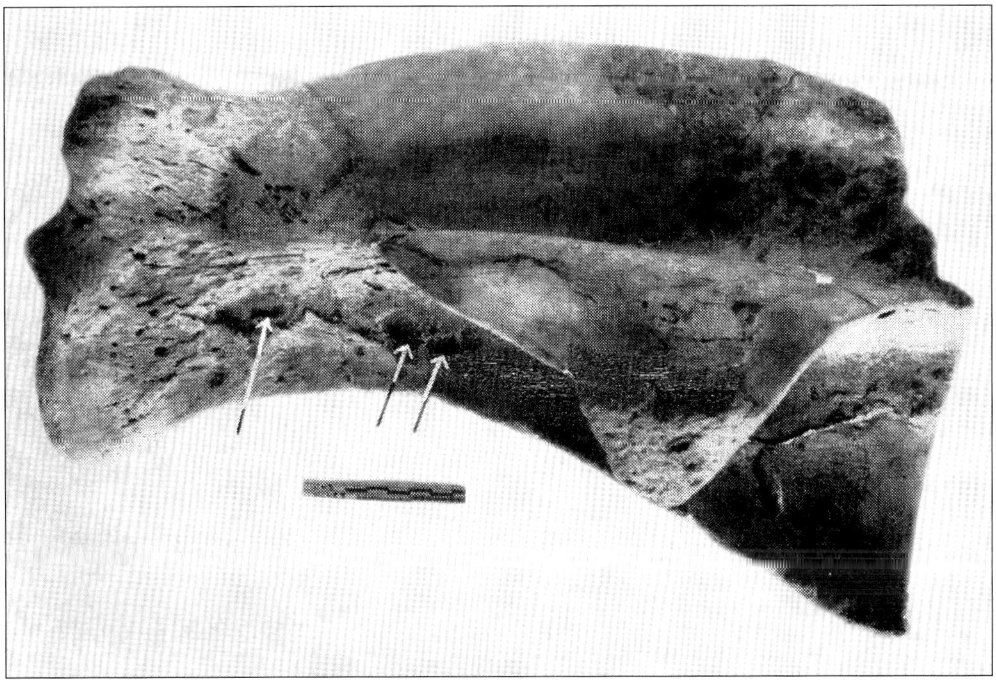

■ Shoulder-blade of an Indian rhinoceros from Langhnaj, Gujarat. The arrows show pits indicating its use as an anvil.
From Sankalia, 1965, plate XIII (1).
© Deccan College, Pune.

vertebra.[17] Though the site had a diverse array of animal species, the contribution of rats/rodents along with possibly the rhinoceros, elephant and hippopotamus to the food economy was considered doubtful.[18] However, that these big mammals were exploited is evident from the fact that the biggest cluster of their remains comes from the Eastern Area or what the excavators called the "butchering complex". The carcasses of large mammals, thus, seem to have been scavenged to obtain meat, and the bones used for fashioning tools.[19]

Moving west, Gujarat is another state where the mesolithic/microlithic culture is well understood on the basis of excavations. The sites of the region yielding traces of the rhinoceros include Khaksar,[20] Valotri[21] and Kanewal.[22] These are microlith-bearing sites which are likely to be late since they also contain chalcolithic pottery.

The most telling evidence, however, comes from Langhnaj[23] in the northern part of Gujarat, where human exploitation of the animal received considerable attention. Frederick E. Zeuner painstakingly demonstrated how the deliberate pits on the shoulder-blade of the animal at Langhnaj indicated its use as an anvil for making microlithic tools. Additionally, from the predominance of game animals in the food debris at the site, he inferred that the economy must have been largely dependent on them. Whether the rhinoceros was a possible inclusion in the diet is a matter of speculation, since he mentions the animal as the most "remarkable" of the game animals at the site.[24] Similarly, Juliet Clutton-Brock, in her examination of the faunal remains at Langhnaj, concluded that all species except the mongoose and the wolf were part of the food economy.[25] In this, was she clearly reinforcing what Zeuner had ambiguously hinted at 13 years earlier?

In view of all this evidence, what seems indisputable is that the economic exploitation of this mega herbivore hearkens back to ancient times. While there is little doubt regarding the use of its bones for fashioning tools, what we still await is direct evidence for its meat being consumed. The possibility however, cannot be precluded if we push our evidence a little more. For instance, a point that deserves some thought is that if the technological level of the mesolithic population enabled

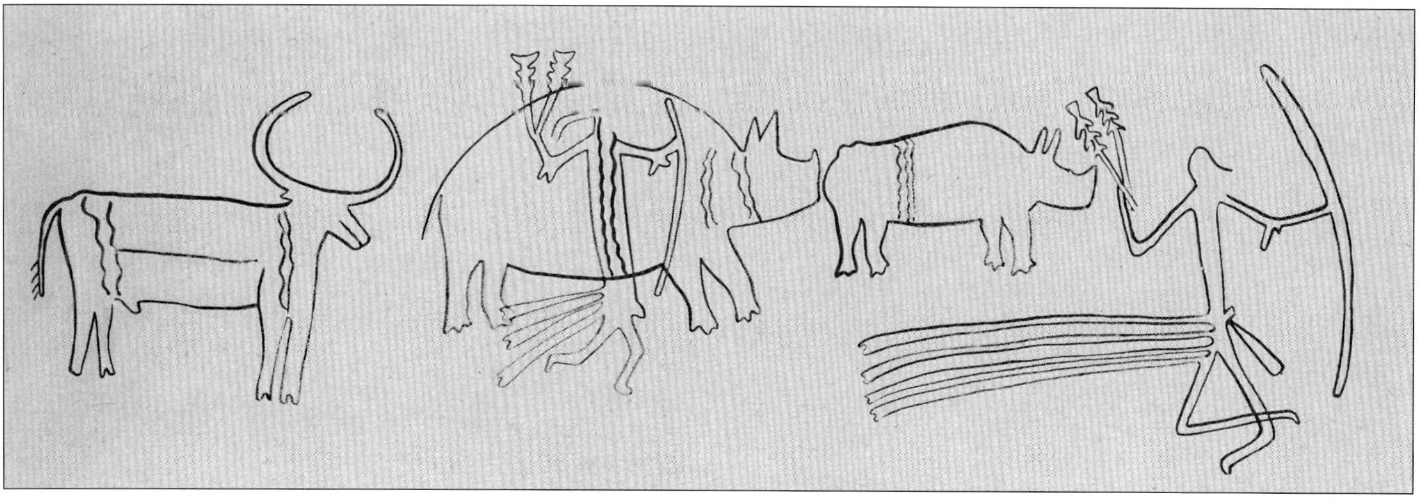

■ Mesolithic representation of two rhinos and a bull, overlaid by the figures of two hunters with bows and arrows, Chaturbhujnath Nala.
From Neumayer, 2013, p. 138.
© Oxford University Press.

the utilization of the bones of a dead rhino (scavenged or hunted), then the same technological level would have led them to utilize its meat as well. It is difficult to believe that a large mountain of meat offered by a rhino would be easily foregone. Despite its hard exterior, a rhino is not so difficult to hunt. In fact, notwithstanding its tough appearance, rhino hide is known to be quite tender at places, making the animal far more vulnerable than it looks.[26] Reservations regarding the use of the flesh of big mammals for dietary purposes, thus, need to be examined from a broader perspective.

The Visual Record

Compelling evidence comes when we turn to the rock shelters of India which house some of the earliest faunal imageries telling us of hunts and animals hunted. Though the chronology of these paintings is problematic, based on stylistic considerations, there have been attempts to broadly assign them to cultural periods.[27] It is to these early visual portrayals that we now turn, in order to take our story forward.

The importance of these representations lies in the fact that they are, perhaps, the first pictorial pointers we have to human interactions with animals. A crucial observation is that not all wild animals are depicted. The subjects chosen mostly seem to be the outcome of a conscious selection that can perhaps be explained by human preoccupation with animals that mattered the most, or that formed a part of the diet, suggesting that early artists were mainly interested in edible animals. Others such as the felines were represented only because they inspired awe.[28]

It is significant, then, that the rhinoceros captured the artist's imagination from early times. Erwin Neumayer estimates that about 30 rhino figures feature in the rock art of the subcontinent, and most of these occur in and around Bhimbetka, at Chaturbhujnath Nala, close to the Chambal valley (both in Madhya Pradesh), and at sites in the Kaimur range above the valley of the Son river, in the Mirzapur region of Uttar Pradesh.[29] The popularity of the animal as a subject of portrayal has often been underlined,[30] and so has the fact that in most depictions it is shown being hunted.[31] While we are conscious of the multiple ways of perceiving such depictions, it seems reasonable to argue that at least some of them can possibly be seen as expressions of experience in actively hunting this big animal. It remains to be asked, then: what became of it after it was hunted? Logic permits us to infer that, once overpowered, the animal would be put to use in its entirety.

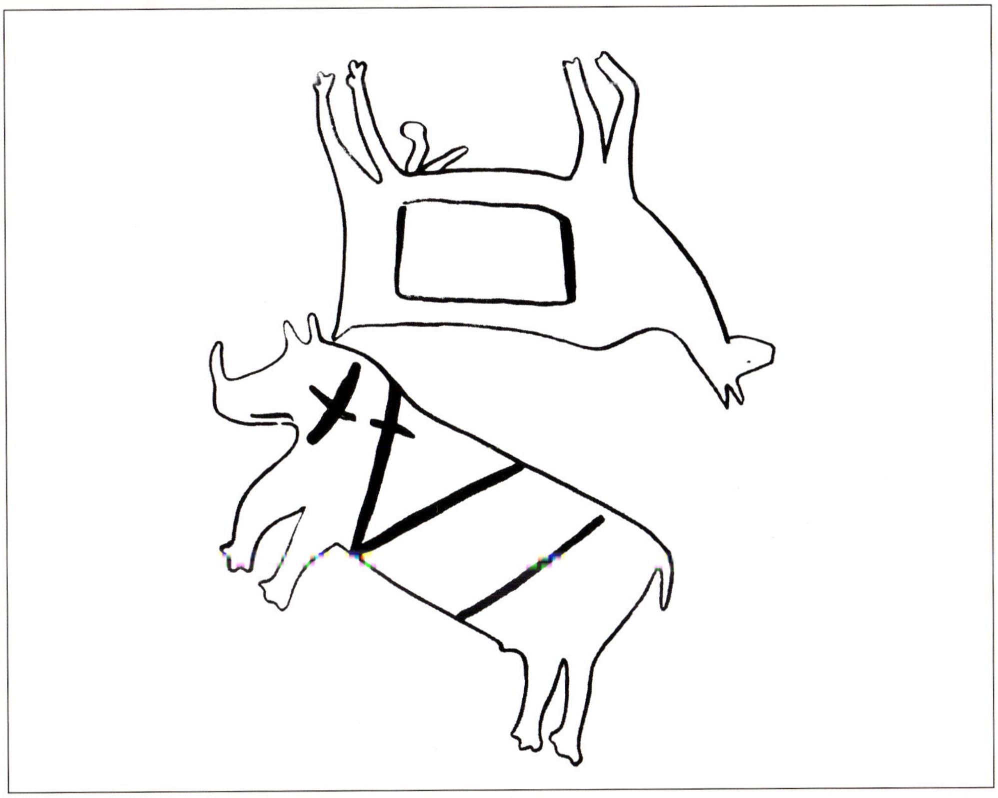

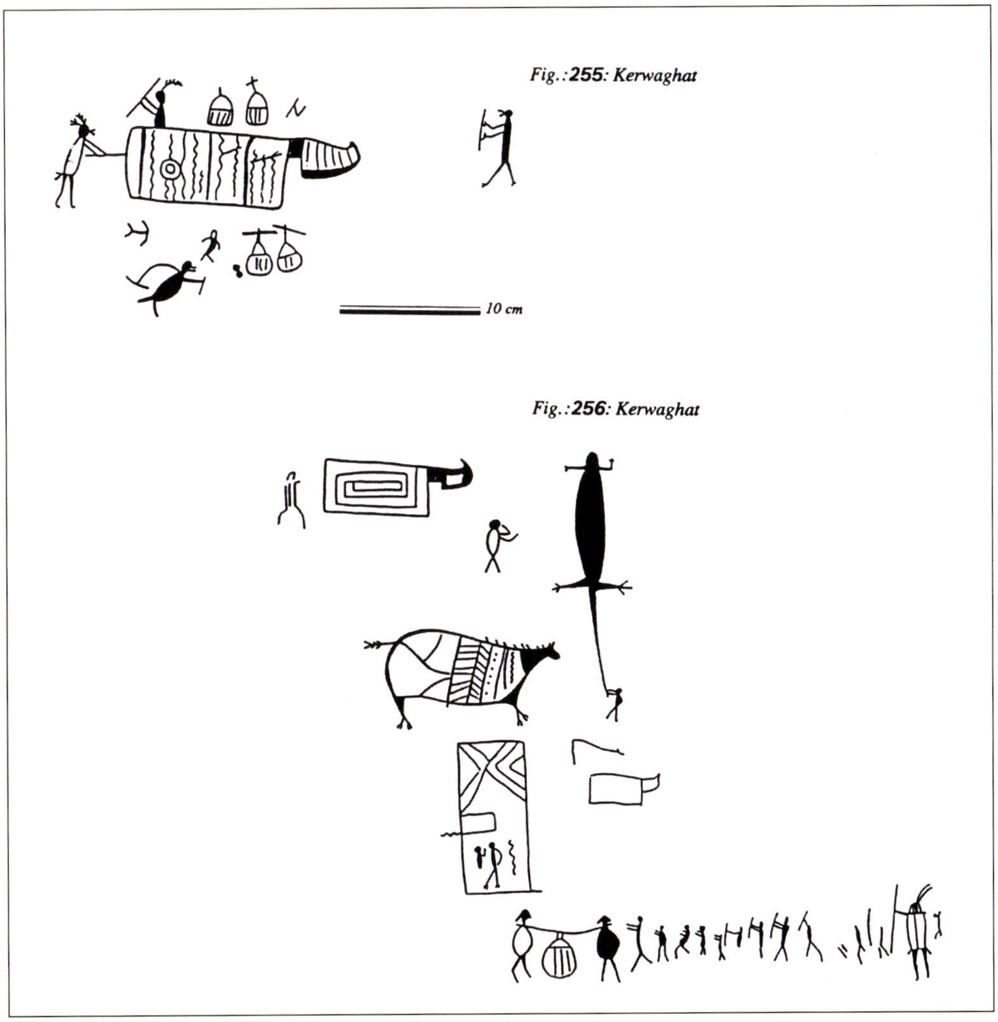

▪ The mesolithic painting represented here, showing a rhinoceros and an antelope, is found on a ceiling at Chaturbhujnath Nala.

From Neumayer, 2013, p. 138.

▪ Stylized rhino depictions in mesolithic (?) rock paintings at Kerwaghat, Uttar Pradesh.

From Neumayer, 1993, p. 115.

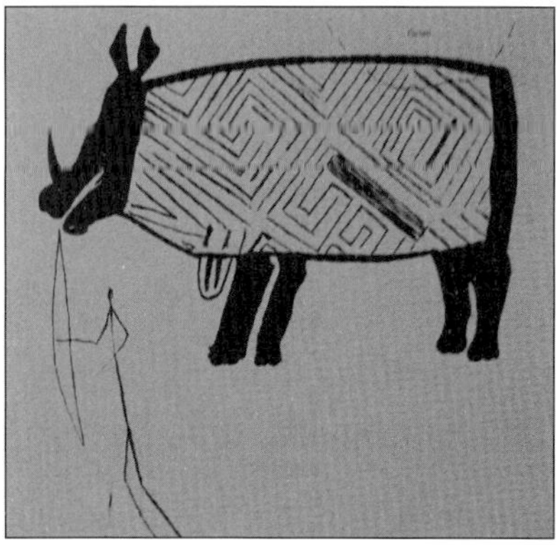

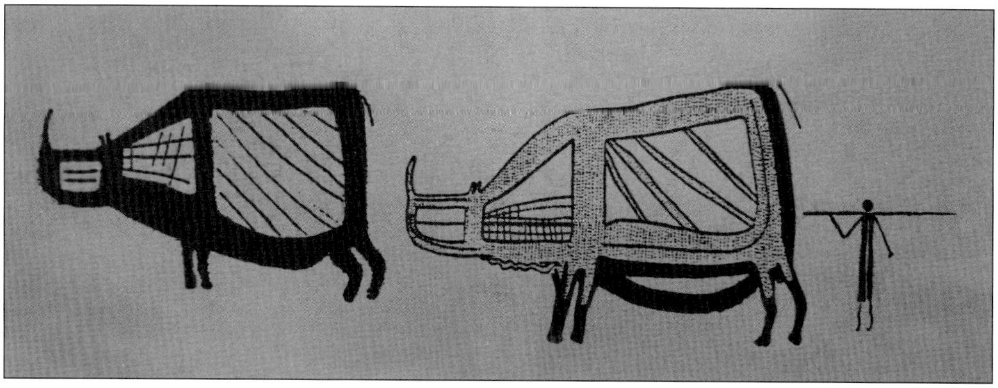

▪ Mesolithic representation from Firengi, Madhya Pradesh, showing an intricately patterned rhinoceros juxtaposed against a stick-like hunter.
From Neumayer, 2013, p. 136.
© Oxford University Press.

▪ Drawing of mesolithic representations of the rhinoceros in a rock painting from Urden, Madhya Pradesh. Juxtaposing stick-like human figures against the large frame of the animal conveyed the dichotomy between human and animal power.
From Neumayer, 2013, p. 136.
© Oxford University Press.

A particularly revealing portrayal comes from Kerwaghat in the Mirzapur district.[32] Presumably belonging to the mesolithic period, it is perhaps the only representation of its kind, and shows the butchering of a large stylized rhinoceros close to what appear to be huts. The human figures seem to have brought vessels and containers to carry the meat away from the site. The next illustration shows two stylized rhinos of varying sizes—the smaller one seems to have been left incomplete. Equally significant is the depiction of the unfinished rhinoceros figure next to what looks like a doorway. While it may be presumptuous to assign conclusive meanings to the portrayal itself, it certainly does not seem far-fetched to visualize the presence of the animal in the vicinity of human settlements in the region during this period.

Other than this, several extensive scenes, also attributed to the mesolithic period, show rhino hunts where the animal is being confronted or pursued, often with dangerous consequences for the hunter himself. For instance, we have the spectacular pictorial narrative from the Ghormangar rock shelter mentioned earlier, where a wounded one-horned rhino is surrounded by 11 hunters who confront it with multi-barbed harpoons that appear to have wooden shafts and stone blades. The animal has been pierced in different parts of the body—on the back, on the hip, and below the knee. Its open mouth communicates its anger at being attacked as well as its anguished plight because of the wounds inflicted. While the swift attacking positions of the hunters effectively convey their desire to overpower the quarry, it also seems to suggest that hunting accidents, where hunters became victims of their chosen preys were anticipated. The perils of hunting are unambiguously conveyed as one hunter is tossed in the air by the horn of the rhino.[33]

There are graphic representations of human attempts at asserting dominance over nature; yet in most cases, what is also unambiguously impressed upon the spectator is the sheer enormity of the task at hand. This is perhaps best illustrated by the contrast in size between the forces pitted against each other. The stick-like human figures seem no match for their formidable adversaries, yet they are recurrently juxtaposed.

The plight of the man caught between a charging rhinoceros and a big boar in a painting at Jaora, in Madhya Pradesh, the largest rock art centre in the country, not only conveys the perils of encounters with the one-horned giant, but also qualifies as one of the best animal portrayals in Indian mesolithic art.[34] Armed with a long spear, the figure seems to be on the run with the rhinoceros close at its heels. As for the portrayal of the animal itself, it is a fairly accurate one showing its bulky body, stumpy legs, folds of the skin around its neck, a prominent horn, and a tiny tail.

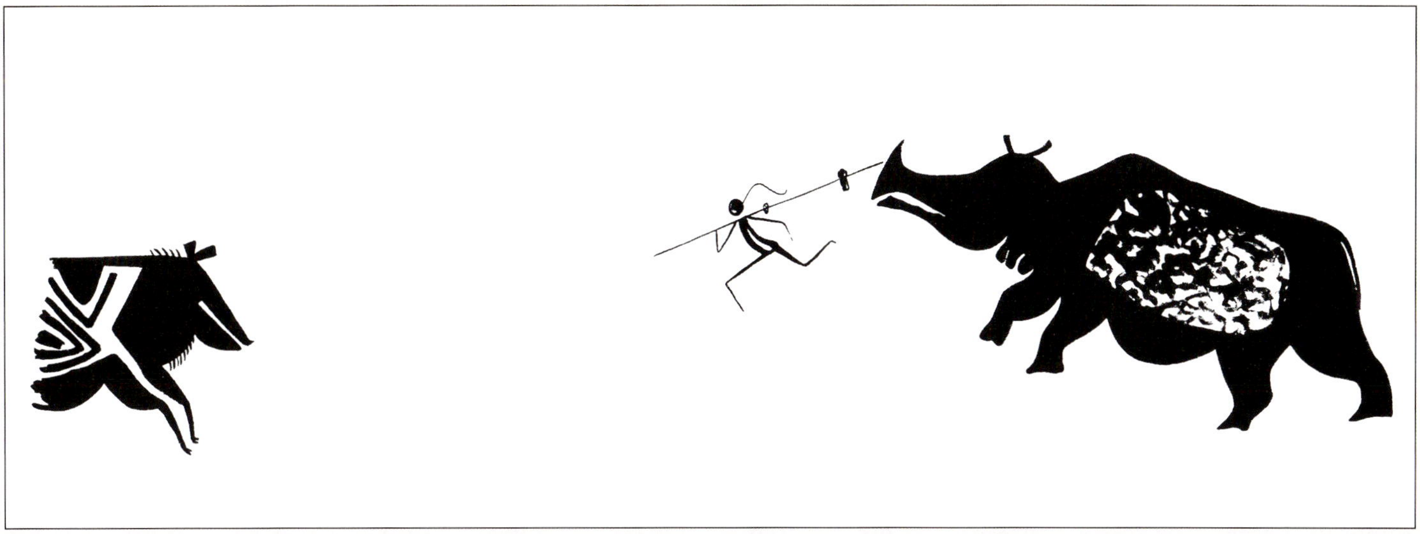

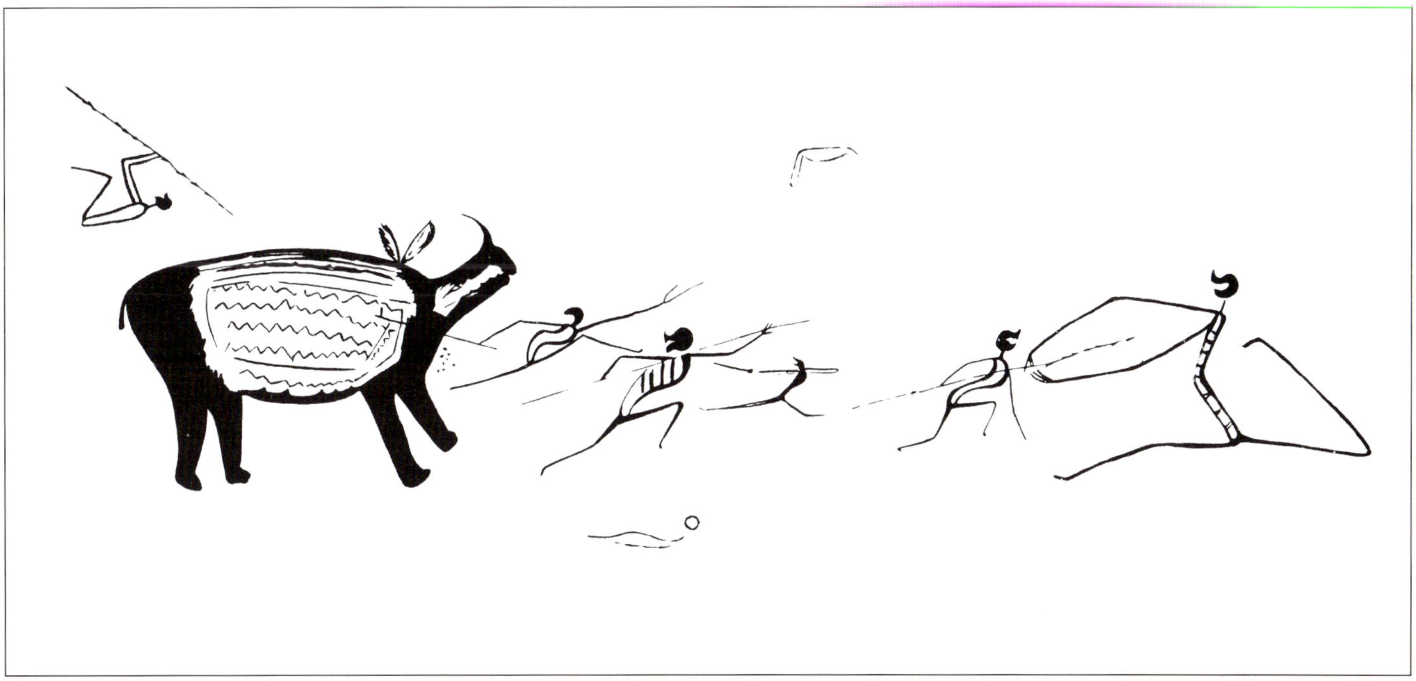

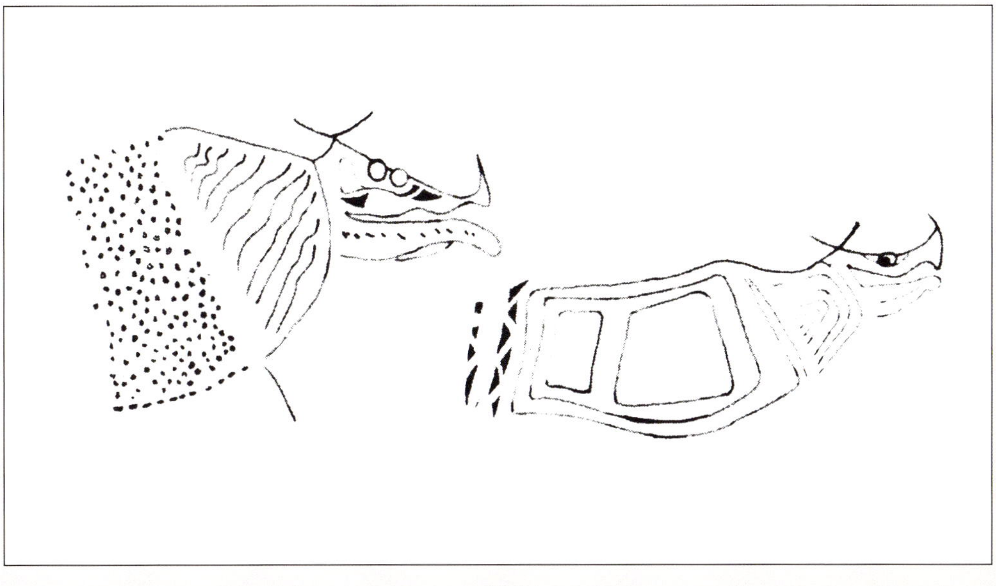

■ Drawing of a mesolithic rock painting from Jaora, Madhya Pradesh, depicting a man caught between a rhinoceros and a wild boar.

From Neumayer, 2013, p. 137.

■ Drawing of a mesolithic rock painting from Bhimbetka, Madhya Pradesh, showing a rhino hunt scene.

From Neumayer, 2013, p. 137.

■ Finely drawn mesolithic rhino representations in a rock painting from Kathotia, Madhya Pradesh.

From Neumayer, 2013, p. 134.

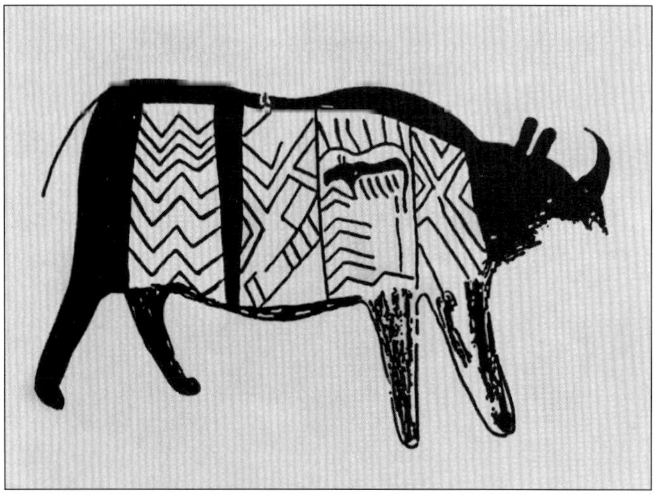

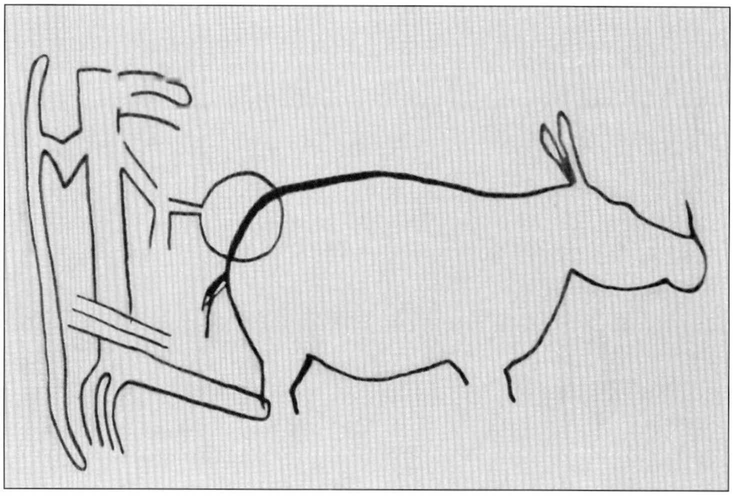

▪ Drawing of a mesolithic (?) rock painting representing the animal from Gelpur, Rajasthan.
From Neumayer, 2013, p. 136.
© Oxford University Press.

▪ Mesolithic representation from Dekan-Matschi Kalla, Madhya Pradesh. The rhino figure is overlaid by that of a human holding a bow.
From Neumayer, 2013, p. 138.
© Oxford University Press.

At Bhimbetka, the animal is surrounded by six hunters holding spears probably tipped with microliths. The massiveness of the creature with its sharp horn, open mouth, raised ears, and twisted tail, is again in vivid contrast to the stick-shaped and small human figures (one of which has been tossed in the air) confronting it.[35] Despite the arduous nature of the enterprise, one of the hunters has managed to strike the big animal in its neck, causing it to bleed.

Though encounters are the most common theme, it is not conflict alone which is depicted. There are also poignant and compassionate glimpses of a mother and baby rhinoceros which captured the attention of early artists, for instance, in the Mithaiya Mand rock shelter of the Kaimur region.[36]

By and large, the paintings divulge an association that unquestionably transcends mere acquaintance. One can think for instance, of the X-ray depictions common in mesolithic art though not confined to it. Very often, these are anatomically correct indicating the intestines, heart, lungs and digestive tract.[37] From Kathotia (Madhya Pradesh), we have fine depictions of the animal assigned to the earliest stylistic group of mesolithic rock art. The eyes, rarely shown in mesolithic portrayals, are shown on the profiled head of the larger rhino.[38] Though not always accurate to the last detail, such close observation of a formidable animal can perhaps only be explained by the fact that it was hunted frequently enough. The fidelity of the portrayals, in most cases, leaves no doubt that they must have been adequately seen around to have entrenched themselves in human consciousness which then sought to express itself through tangible portrayals.

As far as prehistoric rock portrayals of the unicorn are concerned, most seem to be quite explicit in their depictions of the animals, yet there are some which call for attention. One wonders, for instance, at the rather abstract rendering of the rhinoceros in a mesolithic context at Tarsang in Gujarat.[39] To reflect a little on the painting itself, it is easy to see that the peculiar portrayal of the animal dominates the scene. That it is indeed a rhinoceros can be argued on the basis of the depiction of a horn-like projection. Described as a hunting scene, what is noteworthy are the two figures who appear to be holding shield-like objects possibly in their defence. What is intriguing, however, is the absence of any weapons of offence. The animal seems upturned, and the human figures in motion. Evidently, the animal was known, yet chosen to be depicted with an element of abstraction. We can be far from certain about the implications of such garbled representations. The reasons could

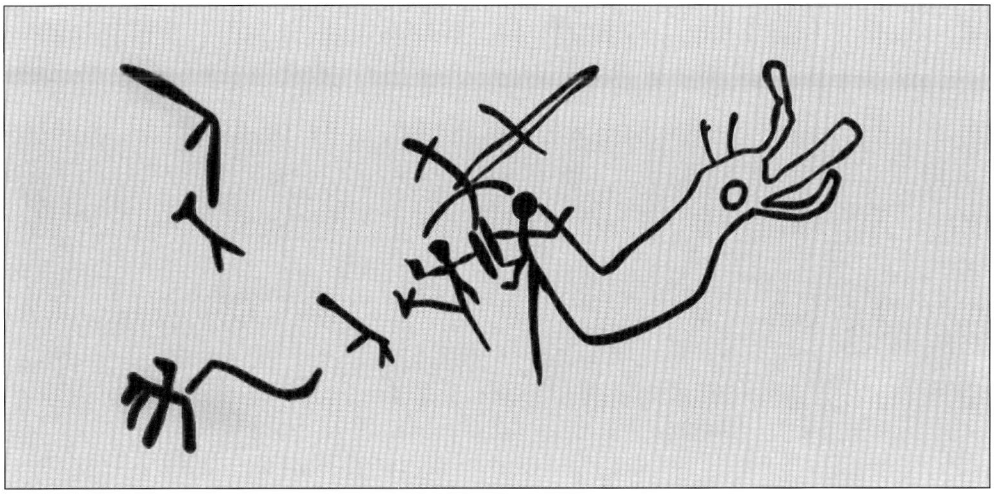

■ Mesolithic hunting scene from Tarsang, with an abstract rendering of the rhino. *From Sonawane, 2002, p. 71.*

be stylistic as well as symbolic, but the fact that they occur gives us a sense of the parallel existence of an imagined world.

Later Periods

Rhinos gradually seem to fade out from the rock art of agriculturalists and herdsmen. An immediate line of reasoning may suggest that the animals were not seen around as often, or even if they were, they no longer served as subjects of portrayal. However, the latter rather than the former seems to be a more plausible explanation for the ebbing out of the animal from the creative canvas of early rock artists. It may perhaps be pertinent to mention here that, broadly speaking, later rock paintings are thematically and stylistically different from their earlier counterparts. The most striking new theme is war, scenes of which pervade the shelter walls.[40] Battle scenes, with riders on elephants and horses, seem to dominate later rock art imageries. Exceptions featuring our protagonist are nevertheless seen, and we shall return to them in the next chapter. Before that, we must collate the faunal evidence that testifies to the presence of the animal in other cultural contexts in the subcontinent.

Actual remains of the mega herbivore were retrieved from neolithic Chirand in Bihar.[41] Ecologically, what is even more significant is the evidence telling us about the presence of the rhinoceros in peninsular and south India. While Paiyampalli[42] in North Arcot district of Tamil Nadu yielded rhino bones in a neolithic context, Nevasa[43] and Inamgaon[44] in Maharashtra testified to the presence of the animal in chalcolithic contexts. Plausibly, then, the environment during the neolithic-chalcolithic period, at least in pockets of peninsular India, must have been conducive for the survival of large mammals like the rhinoceros.

Interspersed as these traces are, they show that even before the advent of the written word, ancient India chronicled its encounters and interactions with the mega herbivore, and that this interface continued to be documented in different ways till literary records help take our story forward.

CHAPTER 3

A Search through Antiquity

Shibani Bose

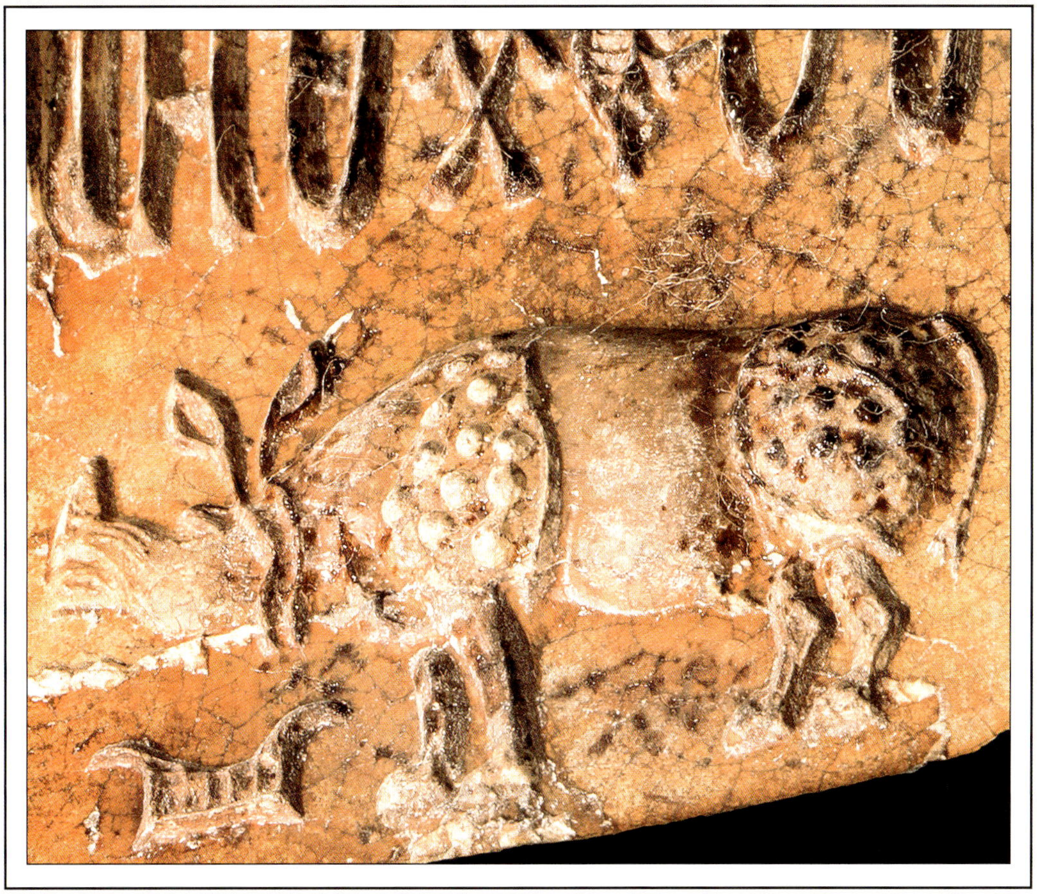

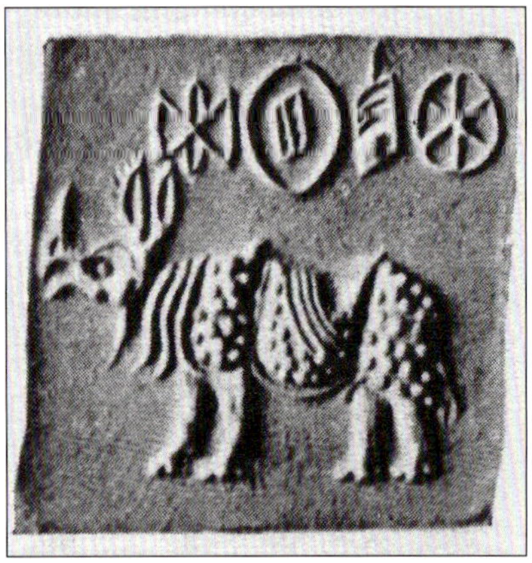

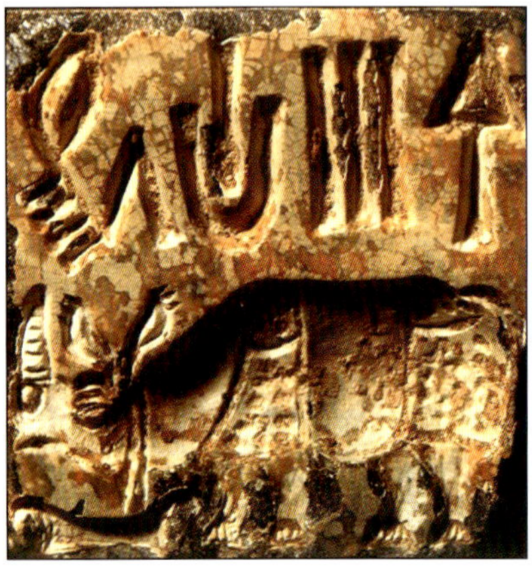

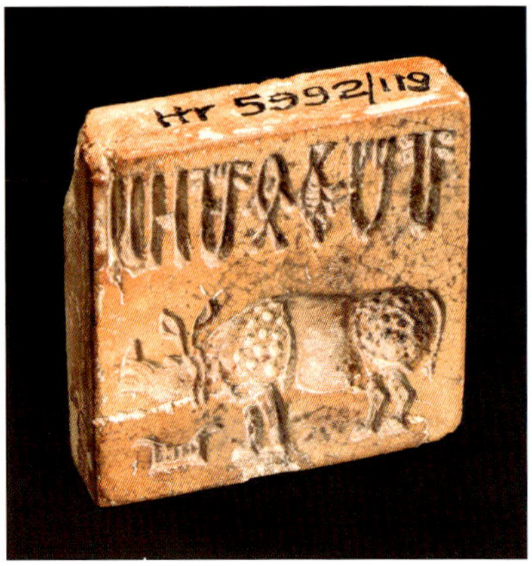

▪ Indus rhino seal without trough.
Courtesy ASI.

▪ A seal from Mohenjodaro showing a rhinoceros with a trough in front. Note the folds on the skin with tubercles and the prominent horn.
Courtesy ASI (after Joshi and Parpola, 1987: 381).

▪ *A Harappan seal showing a fairly well-delineated rhinoceros with trough.*
Courtesy National Museum, New Delhi.

In the next episode of our narrative, we first approach the Indus civilization (circa 2600–1900 BCE)—the first urban civilization of the subcontinent. As is widely known, the pictographic Indus script is yet to be deciphered. Still, the civilization has much to divulge about the rhinoceros through the faunal and visual records it has left behind. As we approach the early historical period, the extent of the visual and archaeological evidence diminishes; however, the rich textual sources provide ample compensation, as we shall see.

Evidence from Indus Sites

Traces in the faunal collections from Harappan sites demonstrate the presence of the rhinoceros at the sites of Nausharo[1] in Baluchistan, Amri[2] in Sindh, and Harappa[3] in Punjab (all in Pakistan), and the Indian sites of Kalibangan[4] and Karanpura[5] in Rajasthan as well as several in Gujarat such as Surkotada,[6] Shikarpur,[7] Kuntasi,[8] Lothal,[9] Khanpur[10] and possibly also Oriyo Timbo.[11] What can safely be inferred on the basis of these records is the presence of the moisture-loving rhinoceros in protohistoric times in areas now known to be arid and semi-arid.

The factors that made for favourable rhino habitats in areas where the animal is now extinct invite some thought about past landscapes. How, for instance, do we reconcile the large amount of evidence pointing to a dry climate during the pre-Harappan and Harappan phases[12] with approaches that have considered the presence of the rhinoceros as an indicator of a humid Harappan environment?[13]

The answer perhaps lies in the ecology of the animal itself. Given the fact that the majority of Harappan settlements flourished in alluvial terrains created by mighty rivers,[14] it is not difficult to envision marshy-swampy patches of these tracts suitable for the one-horned rhino, known to be the most aquatic of the five rhino species. For instance, L.C. Rookmaaker, who has extensively studied the historical distribution of the species in the subcontinent, maintains that it is possible that the animals were encountered near Indus settlements either in the lower part of the valley or perhaps slightly more northwards where Harappa is located.[15]

What nevertheless needs to be explained is the presence, and in cases like Kalibangan, the profusion, of rhinoceros remains at some of the major urban sites of the civilization. Not just bones, even the Harappan visual archive consisting of seals, terracottas and copper tablets seems to suggest that the animal was seen around frequently. If we consider the argument that the bones of animals not usually domesticated or not normally eaten are likely to be rare in archaeological deposits,[16] then Harappan testimonies regarding the presence of the animal along with its evident popularity as a subject of portrayal raise numerous possibilities that must be explored. These range from the exploitation of the animal for its meat and other parts to the role of the animal in the religious life of the people.

Even when we consider the representations of the animal, the qualitative details and graphic fidelity suggest a close familiarity with the subject, which seems remarkable in the context of a civilization known for its urban character. In the terracotta figurines at Mohenjodaro, for instance, the wrinkled skin is realistically portrayed by hatching or pitting in some cases, while in others, strips of clay are used to show the folds in its hide.[17]

The frequency of representations too suggests that the animal was well known. Its popularity as an object of portrayal is reiterated by the fact that rhino depictions comprised 6.3 per cent of the terracotta collection at Harappa. This figure is

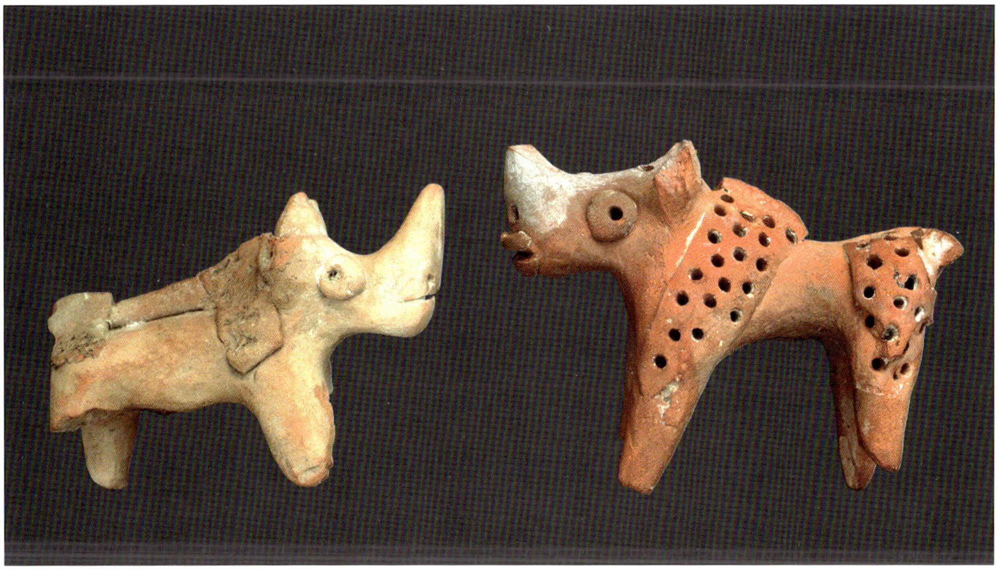

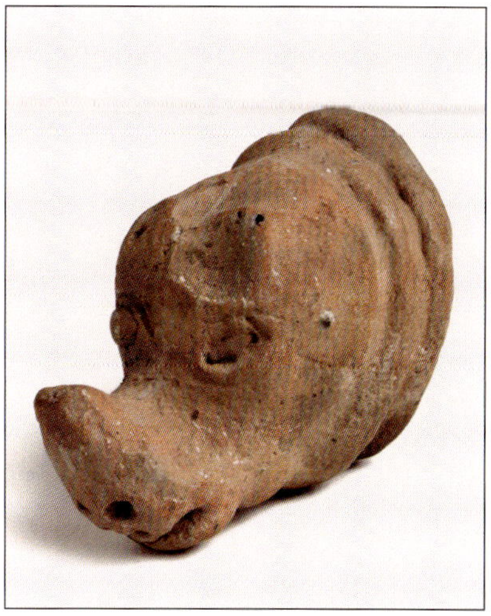

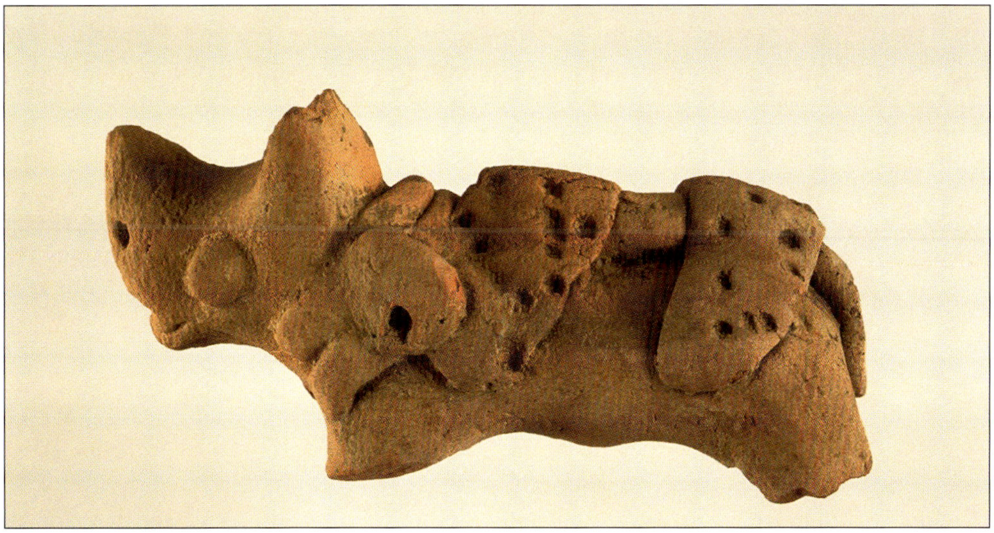

■ Rhinoceros figurines from Harappa. The animal can easily be distinguished because of its prominent horn in both cases. The strips on the back probably suggest the tubercled hide.
http://www.harappa.com/figurines/42.html. Copyright Richard H. Meadow/harappa.com. Courtesy Department of Archaeology and Museums, Government of Pakistan.

■ Rhinoceros figurine from Harappa. Note the animal is strangely portrayed with a collar.
http://www.harappa.com/figurines/43.html. Copyright Richard H. Meadow/harappa.com. Courtesy Department of Archaeology and Museums, Government of Pakistan.

■ Terracotta rhino head, Lothal, 3rd millennium BCE, 4.5 x 6.2 cm.
Courtesy National Museum, New Delhi.

particularly striking since it is more than double the representations of commonly kept animals like sheep and goat.[18]

The animal's appearance on seals leaves no ambiguity regarding the fact that in every case it is the one-horned rhino that is depicted. Elaborating on one such seal, Ernest Mackay noted the well-represented thick hide with the nearly real wrinkles and folds of the skin. Rough excrescences on the skin were indicated in some cases with holes made with a fine drill while in others, hatched lines were employed.[19] The animal was delineated with remarkable faithfulness that even resulted in an exaggeration of detail at times.[20]

Apart from these solitary portrayals, there were more complex ways in which the creature fired the Indus imagination. Curiously, it is sometimes shown standing over a manger-like object. John Marshall was emphatic about the fact that these depictions bore no relation to domestication. His own sense was that these troughs were meant to symbolize food offerings, and that their presence implied that the animals to which these offerings were made, whether in captivity or in the wild, were objects of worship.[21] Equally intriguing is the presence of a collar of two bands in rhinoceros figurines from Harappa.[22]

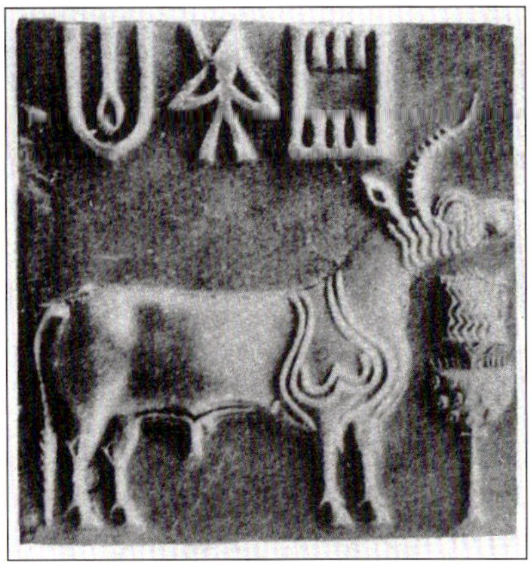

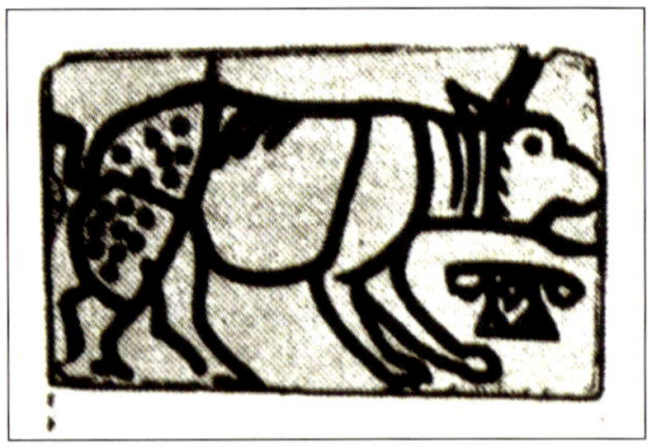

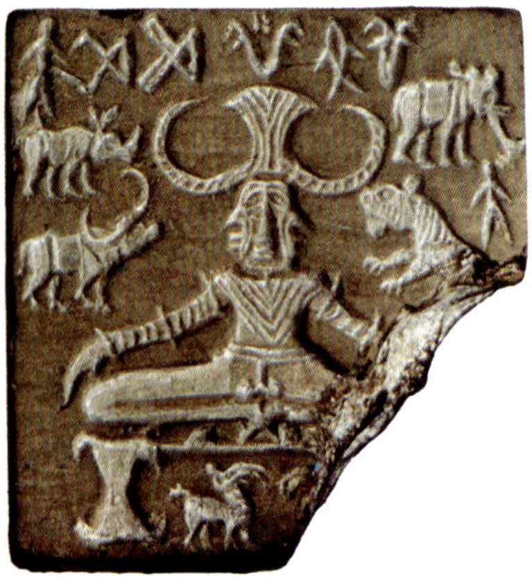

▪ The enigmatic Harappan unicorn: a bovid in profile or a chimera?
Courtesy ASI.

▪ Drawing of an Indus seal with composite animal.
Courtesy ASI (after Marshall 1931).

▪ "Pashupati seal" from Mohenjodaro, circa 2500 BCE.
Courtesy National Museum, New Delhi.

Several Indus seals present a hoofed animal with a single horn. This animal, likely a bovid, probably had two horns but appeared in profile to have one, and was depicted as such. Could a similar profile image of a horned animal be responsible for the conception of the equid unicorn in the European imagination?

Though not a common part of the hybrid compositions found in Indus imagery, one encounters a few representations where the composite figure was the outcome of the amalgamation of the anatomy of a rhinoceros with other animals.[23] Equally significant was the trend of depicting animals moving in single file. This was common on archaic Mesopotamian and Susian seals which showed antelopes and lions, and is also found on some of the seals from both Mohenjodaro and Harappa.[24]

Also relevant to our subject is the celebrated iconographic representation on seal no. 420 at Mohenjodaro. A subject of scholarly scrutiny since the time of its discovery, the seal shows a half-human and half-animal horned figure seated in a meditative posture on a dais. Famously and variously perceived as Pashupati or the prototype of the historic Shiva,[25] the "Mistress of the Animals"[26] or the "Lady of the Beasts",[27] the appellations, male or female, in either case, evidently have to do with the animals surrounding the therianthropic deity. The animals associated with the figure include the elephant, the tiger, the rhinoceros and the buffalo—species indicative of a tree-dotted tall grassland ecology. As far as the delineation of the rhinoceros on the seal is concerned, what is unmistakable in the tiny carving are the leaf-like ears, the folds on the skin, and most explicitly, its one-horned identity. Beneath the dais on which the figure is seated, are what Marshall identified as deer or ibex,[28] though Atre[29] perceived them as composite animals.

If we were to mentally assemble the portrayals of the animal in the context of the Harappan civilization, what immediately comes to mind are intriguing depictions of the creature feeding out of a trough the way tamed animals do, its portrayal with a collar, or even the composite figures combining parts of various animals often with a human face. In the absence of a deciphered script, it may not be possible to access the definite meanings of such portrayals, yet what can be said with a fair amount of conviction is that these had their own symbolic rationale that perhaps reflected a larger worldview in which these animals were used, feared and possibly even deified. Thus, portraying them in ways other than the actual was perhaps part of a vocabulary of shared meanings. It seems reasonable, then, to contend that the animal occupied a significant place in the Indus belief system.

Our search through antiquity for our one-horned protagonist would be incomplete without a mention of its presence in the scintillating hoard of bronzes from Daimabad, a chalcolithic settlement in Ahmadnagar district, Maharashtra, assigned to a Late Harappan context, and chronologically placed between circa 2000 and 1800 BCE. The find, a chance discovery by a farmer, consisted of four bronzes: a chariot pulled by a pair of bulls, an elephant, a rhinoceros and a buffalo. The rhinoceros stands on two horizontal bars placed on the axles of two pairs of wheels. The folds of the skin on the animal's body are stylistically depicted, with those on the back and the belly forming a rectangle. The ears are small, but the muzzle is too elongated for that of a rhino. The tiny horn on the tip of the snout, however, is a definite marker of its identity.[30] Given their bulk and magnitude, these bronze figurines do not seem to have been toys but may possibly have had a religious or ritualistic significance.

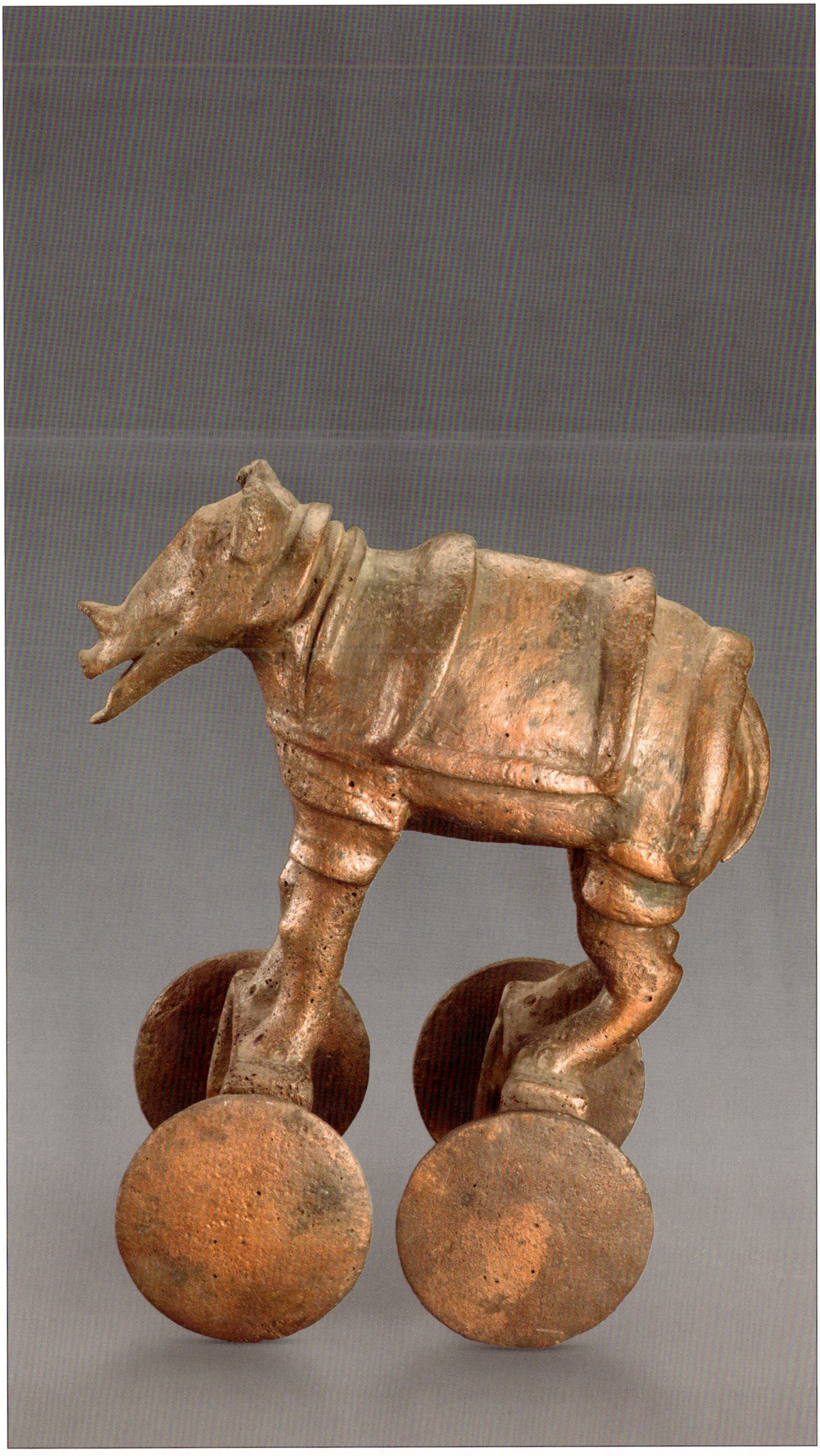

■ Bronze rhino from Daimabad, Maharashtra, circa 2000 and 1800 BCE. Set on wheels, this rhino sports prominent folds on the skin, a short horn and a rather elongated mouth. *Courtesy National Museum, New Delhi. Acc. No. 74.77/3.*

As we travel along gathering tangible as well as intangible traces of the animal, it may also not be out of context to point out how archaeological excavations have demonstrated that the animal's popularity transcended spatial and temporal barriers, by drawing attention to distant sites that have yielded Harappan or Harappan-influenced seals with rhino depictions. At the Sumerian site of Tell Asmar in Iraq, for instance, it is on a cylinder seal of glazed steatite that one encounters the armoured giant in the company of an elephant with a gharial above. The peculiarities of design as well as the subject show such resemblances to the seals of the Indus region that its Indian origin is considered certain. The setting of the ears of the rhinoceros on two little stems is reminiscent of its depiction on Indus seals.[31]

Similarly, a square-shaped Indus seal depicting a rhinoceros along with Indus pictograms also comes from the Harappan settlement of Shortughai, located at the confluence of the Kokcha and Amu Darya rivers in northeast Afghanistan. On the basis of the finds, it was asserted that all the artefacts and technology from the first period of Shortughai clearly originated in the Indus region, and that nothing in those levels could be attributed to another culture or civilization.[32]

Trudging along in the endeavour to reconstruct the past of the animal, we are able to unearth more imprints, though intermittent, of the animal in the subcontinent. One is led, for instance, to Dangawada in Madhya Pradesh, where the chalcolithic level yielded a terracotta figurine of the animal.[33] As mentioned in Chapter 2, though rhino depictions in later rock art are far from common, one does encounter them in the chalcolithic paintings at Ramchaja in Raisen district and Deurkothar in Rewa district of Madhya Pradesh.[34] Similarly, we are also led to Kanyadeh in the Chambal valley, to a composition with letters in post-Harappan characters. It shows a rhinoceros trapped and aimed at by a hunter. Though its antiquity is yet to be fixed, this forms the earliest written record in Indian rock art.[35] Varsus, a site yielding Late Harappan material in the Dhulia district of Maharashtra, revealed a terracotta mask of a rhino.[36]

The Early Historical Period

Though rhino depictions persisted, portrayals progressively lessened. As far as the tradition of portraying the animal in art is concerned, Joseph Manuel points out that something went amiss after the Harappans.[37] The paucity of evidence regarding the animal thereafter is intriguing, and the reasons can best be speculated. What is palpable is a steady distancing of the animal from popular imagination which gradually gets captured with imageries of the mightier elephant and the faster horse.[38] Whether this had to do with the regularity with which these animals were encountered, as also their potential to be tamed, controlled and used vis-à-vis the rhino, is worth considering. We have ample testimony to the existence of the rhinoceros outside its present stronghold in areas of the Northeast until not so long ago, and its unfortunate retreat thereafter.[39] What seems certain, then, is that the dwindling fortunes of the animal had little to do with the diminishing of the species itself. Rather than the animal becoming a rarity, factors engineering the shift seem to have had more to do with changing social and cultural perceptions, conditioned perhaps by new forms of human settlement and production, as also the ebb and flow of kingdoms and cultures. It may well be argued that as agriculture and the clearing of forests became important, both horses and elephants found greater use as domesticates for harnessing and other purposes. The rhino, on the other hand, was not encountered or engaged with on a regular basis. Rhinos could be tamed, but had no draught or domestic use.[40]

There is more about the animal that evokes reflection. What is perplexing, for instance, is that, despite being the second-largest land mammal after the elephant, the rhinoceros figures but rarely in early Indian sculpture. Art historian Joachim Bautze observes that the animal is shown mainly in the earlier periods, but its importance could never rival that of the elephant.[41] What is being emphasized here is not an absence of representations, but a gradual petering out of the tradition, particularly as we approach the historical period in India. This seems peculiar in the context of an animal which had enjoyed relative eminence in the visual records of prehistoric and protohistoric India.

It is equally strange that, despite its impressive size, famed ferocity and legendary association with divinities,[42] the rhino never got assimilated in Hindu mythology and iconography as the vehicle or *vahana* of any major deity in the pantheon[43] (the goddess Dhavdi Ma whose temple is at Dhrangadhra in Gujarat has a rhinoceros *vahana*). An attempt to explain this anomaly speculates if an "ugly" animal befits a god.[44] But a quick look at common mounts like the tiny mouse accompanying the elephant-headed Ganesha, or the buffalo with Yama, the god of death, calls for a reconsideration of this view that underlines a charming exterior as a prerequisite for qualifying as a *vahana*. One wonders if this had something to do with the rhino's infrequent encounters with humans as also the fact that the animal was neither domesticated nor ridden, nor did it induce mortal fear of the kind that the big carnivores did.

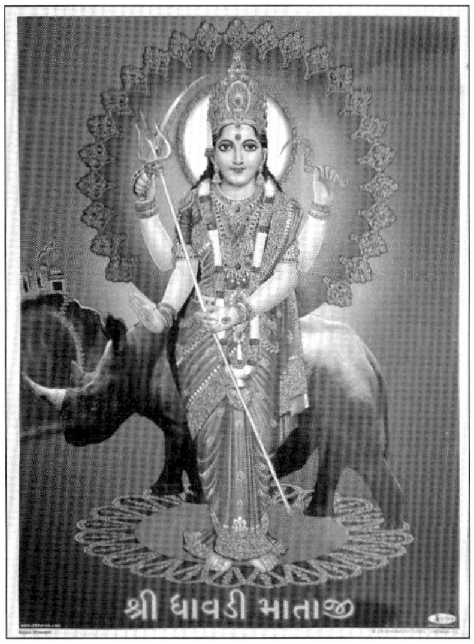

■ Sri Dhavdi Mataji, popular print sold in Dhrangadhra market.
Courtesy Ranmal Jhala, Dhrangadhra.

Having dwelt upon the chequered past of the animal, we return to see what the archaeological record holds for us as we enter the early historical and historical periods in ancient Indian history. The rhinoceros figures in popular imagination during the Northern Black Polished Ware period. The realistic modelling of the terracotta figurine of the animal in deep red ochre paint retrieved from sub-period IB of Period II (circa 600 to the 2nd century BCE) at Kausambi in Allahabad district of Uttar Pradesh clearly indicates the familiarity of the potter with this creature.[45]

We meet our hero again during a survey of the gamut of Mauryan and late Mauryan art objects. Although largely confined to the 3rd and 2nd centuries BCE, some of them are believed to be a century or so later.[46] These include a red and grey soapstone seal dated to about the 3rd century BCE, from Bhita in Allahabad district of Uttar Pradesh, showing a rhinoceros.[47] The mega mammal also figures on a stone disc of the Murtaziganj-Patna group.[48]

It was, however, the sculptors at Sanchi who firmly ensconced the animal in the enduring medium of stone. Amidst the reliefs of the ground balustrade of Stupa 2 (sculptures dating from the last quarter of the 2nd century BCE), we find the rhinoceros. Compared to the care taken in carving the elephant, for instance, amidst a range of other animals, it is not a very successful attempt.[49] What provides a suitable pretext for the presence of the animal on the lower half-medallion on Pillar 24a, is perhaps, the oft-repeated entreaty in Buddhist texts to emulate the solitary behaviour of the rhinoceros.[50]

A lesser-known tryst with the animal is on a coping stone assigned to circa 100 BCE from Mathura, where "a little rhinoceros strides along with stocky stiff legs".[51] The mega herbivore also figures on an *ayagapata* (votive tablet) belonging to the same period at the site.[52] What deserves mention is that the rhinoceros appears in

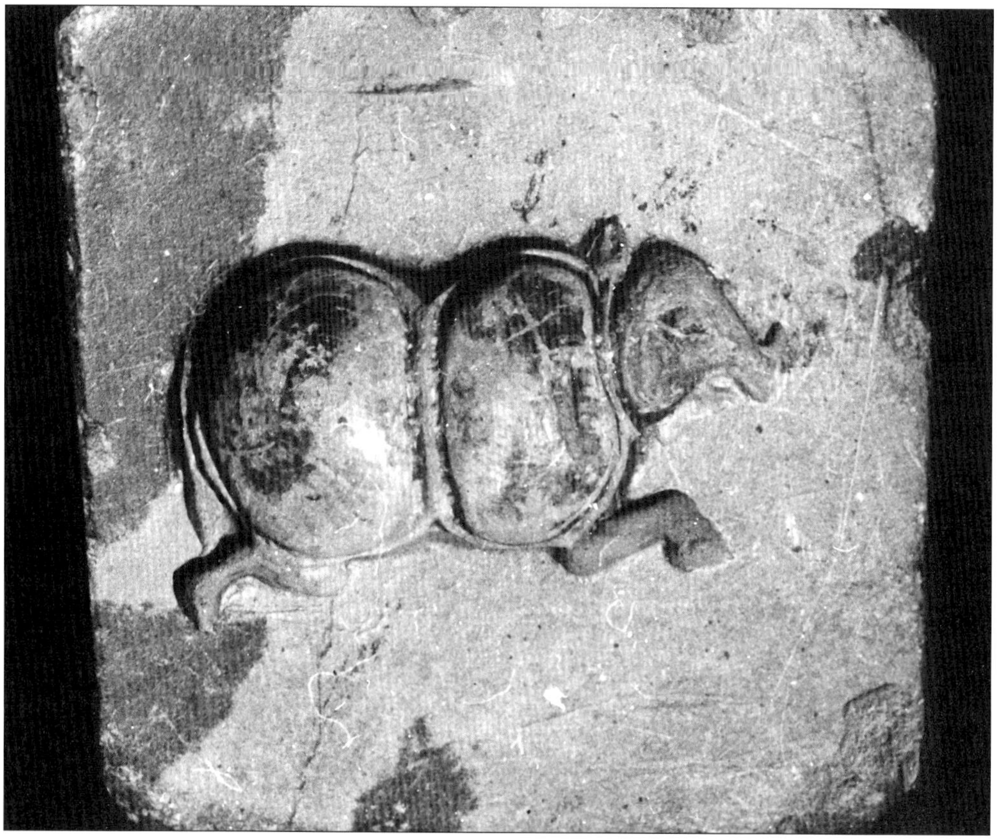

▪ Red and grey soapstone seal from Bhita, Allahabad, circa 3rd century BCE. *Photograph: American Institute of Indian Studies. AIIS: 86.13. Acc No. 84070.*

later Jain iconography as the *chinha* (symbol) of the eleventh Jina, Shreyansnath.

We can also turn to the eastern part of the subcontinent, where the discovery of terracotta plaques depicting the animal amongst the antiquarian finds at Chandraketugarh and Dum Dum, in the North 24 Parganas district of West Bengal, suggests its presence in this region in the early centuries of the Common Era. At Chandraketugarh, approximately 37 kilometres northeast of Kolkata, the terracotta plaque under consideration depicted a standing rhinoceros, chronologically placed between circa 1st century BCE and 1st century CE.[53] This has been assigned to the Kushana period on stylistic grounds. Similarly assigned to the Kushana period, the animal on the plaque from Dum Dum gives away its identity by characteristically displaying a horn on the snout, folds on the skin, and tubercles on the shoulder, thighs and rump. The emphasis on such details makes it a realistic rendition, suggesting the familiarity of the artist with the species.[54]

By far the most outstanding and dramatic representation of the animal in the historical period is encountered on a gold coin issued by the Gupta ruler Kumaragupta I (circa 413/415–455 CE). Classified as the rhinoceros-slayer type, the use of the mega herbivore as a symbol on the coin by the sixth scion of the Gupta dynasty stands out as an innovation against the tiger-slayer type of Samudragupta (circa 350–370 CE) with the legend *vyaghraparakramah*, and the lion-slayer type of Chandragupta II (circa 376–413/415 CE). The obverse of the rhino-slayer coin shows Kumaragupta with a sword drawn in his right hand, mounted on a fully caparisoned horse. The spotlight however, seems to rest on the rhinoceros being pursued by the emperor. As the animal turns its head around to see the imminent danger trailing it, the alarmed look on its face is unmistakable. What also cannot be missed are the attempts to delineate its anatomy.

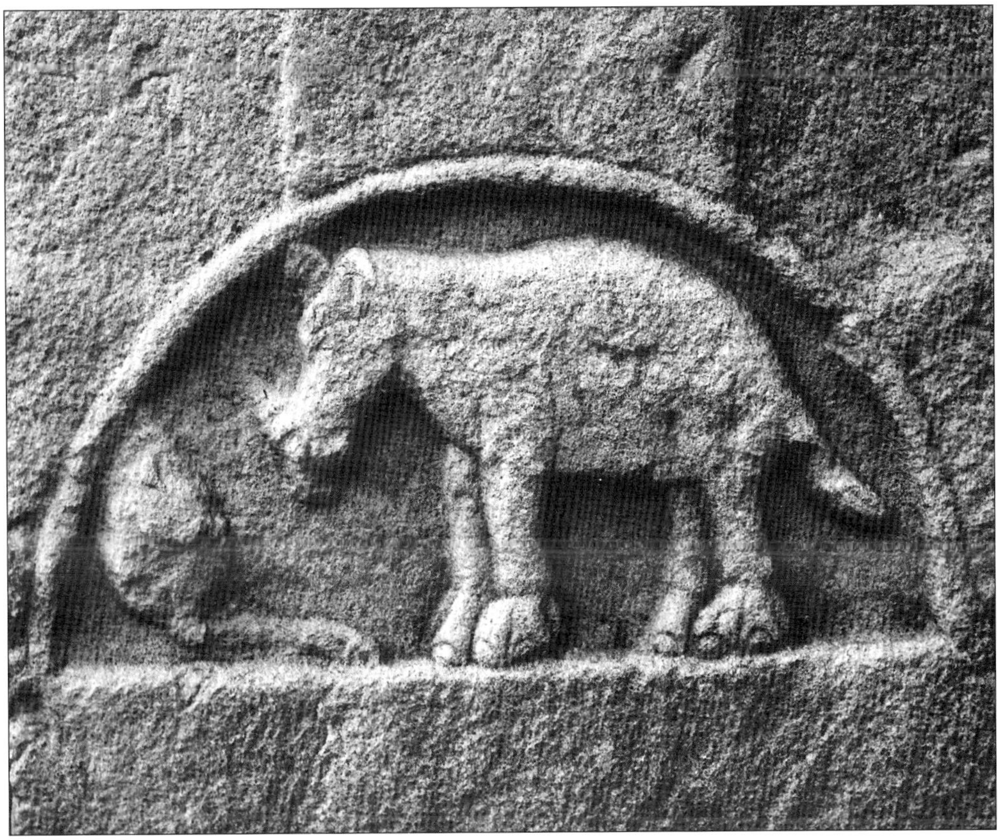

▪ Pillar 24a at Sanchi Stupa 2, medallion with rhino.
Photograph: American Institute of Indian Studies. AIIS: 311.98. Acc. No. 39438.

Pompous legends are used to accentuate the prowess of the ruler. On the obverse is *bharta khadgatrata Kumaragupto jayatyanisam,* while on the reverse is *Sri Mahendrakhadga*. What deserves to be underlined here is the use of the word *khadga* which can be interpreted as a rhinoceros as well as a sword, both of which are seen on the coin. One translation would be: "Ever victorious is the Lord Kumaragupta who is *khadgatrata*, that is, a protector by the sword from the rhinoceros." The legend on the reverse has been interpreted as "Sri Mahendra, the slayer of rhinoceros".[55]

The significance of this coin has been a matter of much deliberation, with suggestions ranging from political and military ones to those that have attributed a religious significance to it. Then there are others that have perceived the coin simply as another of the *mrigaya* or hunting type issued by the Gupta rulers, celebrating their love for big game.

Notwithstanding such interventions, what can perhaps be safely presumed is that the tiger and the lion having already been appropriated by his illustrious predecessors, it was Kumaragupta's quest for a new symbol which led him to the rhinoceros. The sovereign, evidently, seems to have been more ambitious as well as eager to augment the numismatic legacies of his forerunners. The attempt, therefore, was not only to integrate old symbols but also to introduce new ones and the burly unicorn readily offered itself for adoption. The elephant of course would have been a more obvious choice but, given the centrality accorded to the creature, both as a war and as a state animal, it was probably considered prudent to portray it as a mount aiding access to the game rather than as the game itself. It has been argued that although the coins of Kumaragupta depict the rhino, these are aberrations since they are not popular representations. Rather, they are clearly in exaltation of the ruler

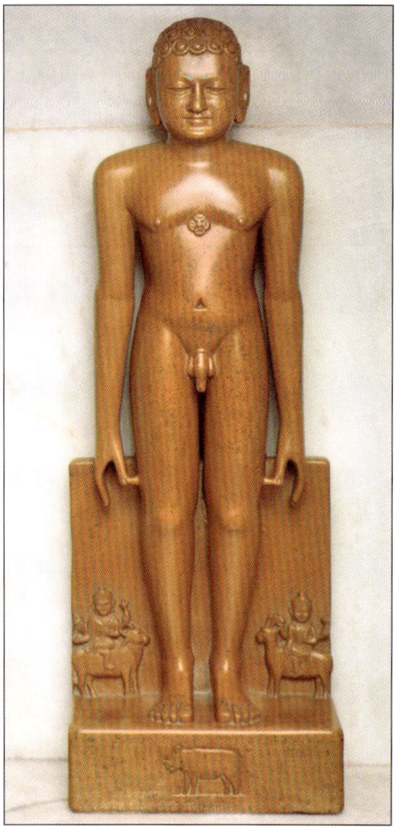

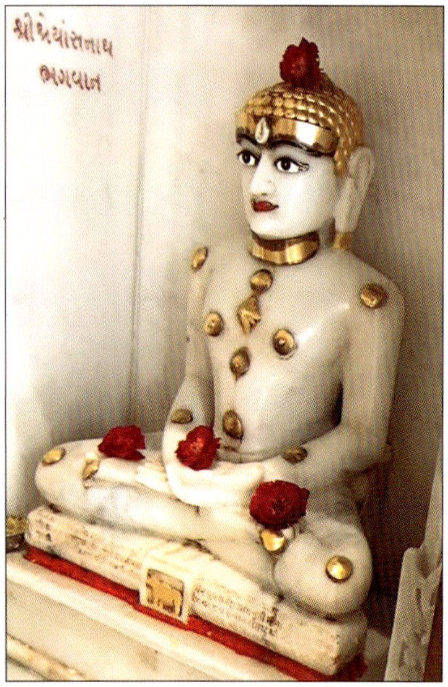

▪ Shreyansnath, the 11th Jain tirthankar, has a rhino for his "chinha" or symbol. The standing statue is that of the Digambara sect of Jainism.
Courtesy Sudhanshu Kasliwal.

▪ Jina Shreyansnath depicted in the Shvetambara sect tradition of Jainism. The symbol remains unchanged—the rhino—though the sect is different.
Courtesy Daivesh N. Shah.

slaying the animal and overpowering its might.[56] While that may be one perspective, the depiction of the animal on the coin should not be merely seen as a device employed to assert the prowess of the ruler vis-à-vis forces to be reckoned with. It is also an indication of the animal having been sufficiently present to be encountered during hunts, at least in the Ganga valley.

Even as late as the 9th and 10th centuries, the rhinoceros found a place in the terracotta assemblage retrieved during excavations at Shyam Sundar Tilla in South Tripura district.[57] This comes as no surprise as, even in the 18th century, northern Bengal and Assam were known to be so rich in rhinoceroses that a French map of India describes that area as "Contrée de Rhinoceros", and late medieval temples in Bengal, approximately from the same period as the map, are decorated with terracotta panels showing rhinoceros hunts (see Chapter 5).[58]

Notwithstanding this evidence, we have to acquiesce that with the dawn of the historical period, there is a gap in the archaeological record. This can primarily be attributed to historical archaeology's preoccupation with prehistory and protohistory. The drying up of archaeological evidence is, however, made up for by the vivid imageries derived from texts and textual traditions. In what follows, we run through some principal texts of ancient India in search of the mega herbivore. These glimpses are merely illustrative and not exhaustive. The intention is not to convey a linear progression from one corpus to the other, but to carve out a narrative of the fortunes of the animal within religious and secular traditions separated in time and space.

Literary Sources

The Vedas provide the earliest glimpses of the human-rhino interface. We tread on firm ground in our search for the armoured giant when we encounter the *khadga*. A pointer to the identity of the animal is that several Vedic passages situate it "in the realm of fierce wild beasts and suggest that its hide is armour-like", an observation that accurately describes the Indian rhino.[59]

Our search for the *khadga*, for instance, leads us to the scene of the *ashvamedha* sacrifice where animals dedicated to different deities are tied to the 21 *yupas* (sacrificial stakes) and in the intermediate spaces. While domestic animals are bound to the stakes, in the spaces between them are confined wild animals including the elephant and the rhinoceros. However, not all these animals are killed, some including the rhinoceros being temporarily confined till the culmination of the ceremony.[60] In the enumeration of the animals assigned to different deities, the rhino is dedicated to the Vishvadevas (all gods) who are worshipped through the Vaishvadeva *homa* (rite).[61]

Though such contexts do not suggest anything more than a ritual significance of the mighty animal, we can perhaps turn to more telling clues for allusions which reflect an interest in the use of the animal. One learns, for instance, about the Vedic use of rhinoceros hide (*khadgakavacha*) in a ritual *dakshina* or priestly gift at the one-day Soma rite known as the Apachiti.[62] The *Sankhayana Shrautasutra* mentions that "the sacrificial fee is a horse-chariot, coated with rhinoceros-hide, covered with tiger fell, with a quiver boar-hide, with a bow-case of panther-hide, drawn by brown horses."[63] Similarly, in a ritual context, the *Jaiminiya Brahmana* stipulates the use of armour of rhinoceros hide.[64]

It is likewise possible to retrieve glimpses of the animal from the corpus of early Buddhist literature based on the Pali *Tipitaka*. Our search for the rhino here, leads

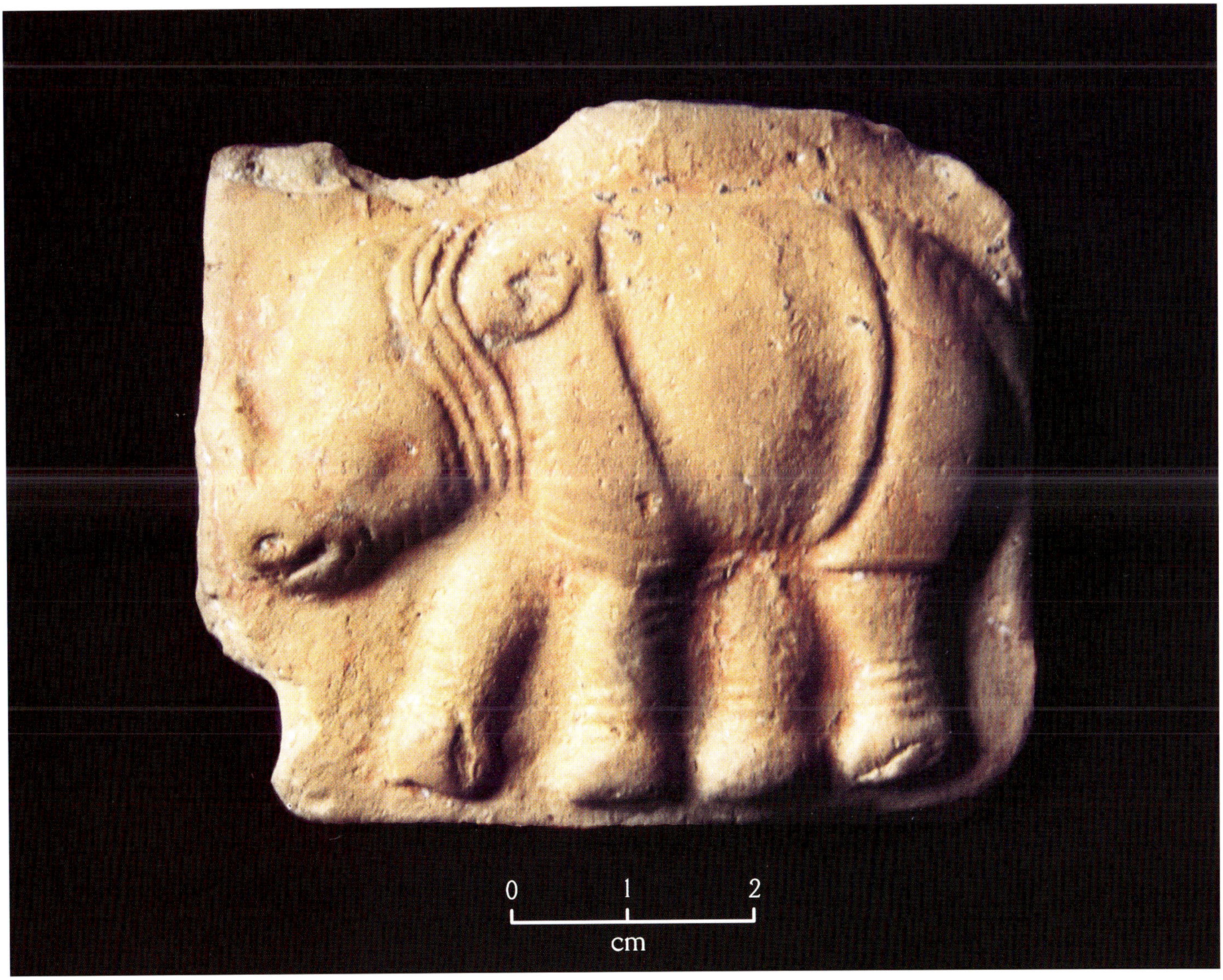

■ Kushana-period terracotta plaque with the image of a one-horned rhino, Chandraketugarh, West Bengal. *Courtesy Archaeological Survey of India.*

us to the *Khaggavisana Sutta* or the rhinoceros sutra of the *Sutta Nipata*, a text which immortalized the animal by extolling its solitary character. The constant refrain *eko care khaggavisanakappo* urges one to wander alone like a rhinoceros.[65] Though allegorical, it is remarkable how this *sutta* shows familiarity with animal behaviour.

In the oft-told Jataka parables narrating the stories of the Buddha's former births, the rhinoceros puts in an appearance in the *Sudhabhojana Jataka* as part of the setting of a hermitage.[66] Similarly, the *Vidhurapandita Jataka* envisions a captivating view of the landscape, rich in fauna including the rhinoceros, when it mentions a magic jewel through which the entire world could be seen.[67] The contexts suffice to convey the ecological sensibilities of ancient India by situating the rhino in environs typically conducive as habitats.

Assigned to the 1st century CE, the narrative in the non-canonical *Milindapanha* centres around a chain of discourses between King Milinda and Nagasena the Elder regarding a number of points of Buddhist doctrine.[68] An interesting allusion occurs to "an elephant hemmed in by rhinoceroses".[69] Though clearly allegorical in the given context, it may be worthwhile to point out that the elephant-rhinoceros

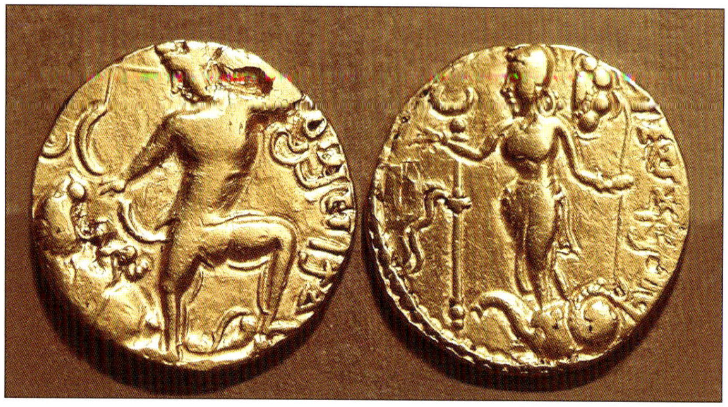

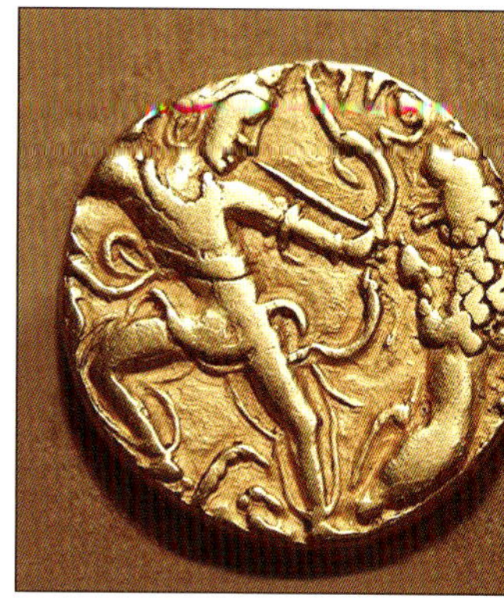

▪ Tiger-slayer type gold coin of Samudragupta (ruled circa 350–370 CE). *From http://coinindia.com/Samudra-4792tigerB-491.10.jpg. Courtesy Indian Museum, Kolkata.*

▪ Lion-slayer type gold coin of Chandragupta II (ruled circa 376–413/415 CE). *From http://coinindia.com/Chandra-4814.2Ev-503.20.jpg. Courtesy Indian Museum, Kolkata.*

animosity (not conclusively proved in the wild) is often referred to in later popular writings.

As in Buddhism, the animal elicits attention because of its solitary character in the Jaina world. The *Kalpasutra*, for instance, extols the fortitude of Mahavira by saying that he was "single and alone like the horn of a rhinoceros".[70] The natural behaviour of the mega herbivore, thus, makes for an image which cuts across traditions.

Significantly, ancient Indian texts suggest that an interest in the skin of the animal, as seen in the Shrautasutras, graduates to an interest in its meat. This, for instance, clearly comes through in the Dharmasutras of Apastamba,[71] Gautama[72] and Baudhayana,[73] which widely forbid the eating of the flesh of "five-nailed" (*pancanakha*) or "five-toed" animals, except for a restricted list comprising the porcupine, hedgehog, monitor lizard, hare, tortoise and very often the rhinoceros. A similar injunction occurs in the *Manava Dharmashastra*.[74] The rhinoceros, though, is a strange inclusion, since it has only three toes on each foot. Nevertheless, if we remind ourselves of the importance assigned to the animal in Vedic ritual contexts, it seems reasonable to argue that this inclusion was possibly an extension of a tradition which had firmly embedded itself in early human consciousness.

Overall, what manifests are attempts to sanction the eating of rhino meat which is further amplified when the Dharma texts unanimously reinforce its preeminence in appeasing ancestors. Apastamba, for instance, tells us that the meat of a rhinoceros (*khadgamamsa*) offered on a rhinoceros skin (*khadgopastarane*) is said to gratify ancestors for an unlimited time.[75] In a similar vein, the *Manava Dharmashastra* spells out the periods for which the flesh of animals offered at rites for ancestors satisfies them, and rhinoceros meat is enlisted amongst the items that are "efficacious in perpetuity".[76]

What clearly emerges from this maze of injunctions and the attempts to explain them is the interest ancient India had in the skin and meat of the rhinoceros. Though the contexts of use in Brahmanical sources are primarily ritual and relate to socio-religious prescriptions, texts like the *Charaka Samhita* and *Sushruta Samhita* (dated to the first half of the 1st millennium CE) and the *Arthashastra* (c. 300 BCE–200 CE) demonstrate how these extended to strategic and medicinal realms as well.

In an eightfold classification in the *Charaka Samhita,* based primarily on feeding habits and habitats, the rhinoceros belongs to the *anupa* class of animals or those which are inhabitants of marshy lands.[77] The meat of the animal is said to impart strength and alleviate *vata dosha* (the body substance derived from the element air). According to Charaka, it is sweet, unctuous (oily), nourishing, beneficial for

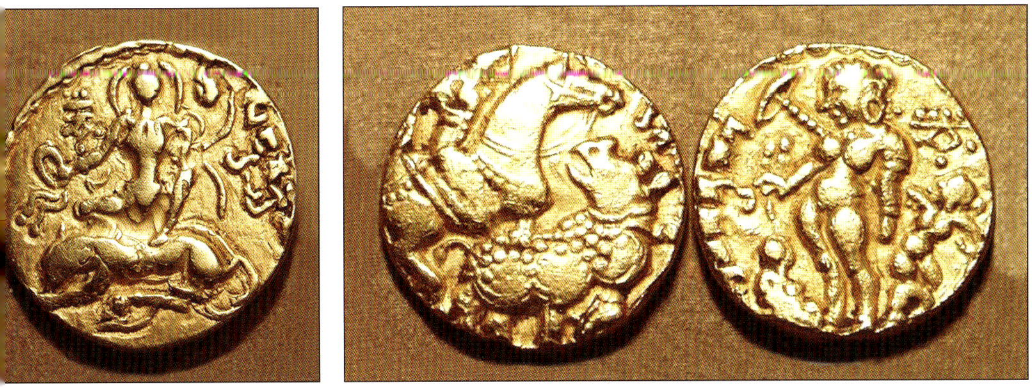

■ Rhinoceros-slayer type gold coin of Kumaragupta I (ruled circa 413/415–455 CE).
From http://coinindia.com/Kumara-Rhino-R8.0-17.jpg. Courtesy Indian Museum, Kolkata.

the complexion, and relieves fatigue.[78] Olivelle contends that strength here may well suggest sexual potency.[79] Elsewhere, the physician is urged to give the well-spiced meat of the animal as a cure for emaciation.[80] Apart from an interest in the flesh of the animal, the text also mentions the use of its horn, stipulating that amulets to be worn by a child be made of the tip of the "right horn" of a living rhinoceros, deer, gayal or bull.[81] The allusion, though ambiguous, does give an inkling of the beliefs associated with the horn of the animal.

In a more complex classificatory system, in the *Sushruta Samhita,* we encounter the *anupa* category again, but here it is divided into five sub-groups where the rhinoceros, along with elephants, boars, buffaloes and other animals, appears amongst those living on banks (*kulachara*).[82] Significantly, the flesh of these animals is said to be sperm-producing and effective in eliminating imbalances in *vata* and *kapha* (the latter *dosha* deriving from the elements of water and earth). While the flesh of these animals is said to be sweet in taste and cooling,[83] the flesh of the rhinoceros specifically is said to have an astringent taste. The meat of the animal is pleasing to ancestors, is sacred, imparts longevity, tends to suppress the discharge of urine, is dry and pacifies *vata* and *kapha*.[84]

Such classificatory systems not only reflect an engagement with the ecology of the animal but also underline the preservation of the lore recommending rhino meat. The *Sundarakanda* of the *Ramayana* recounts how, as part of his quest for Sita, Hanuman explores Ravana's palace, searches the harem as well as the drinking hall with the aroma of delicacies which also include "*vardhranasakas*".[85] J.L. Brockington emphasizes as an oddity this sole occurrence of the rhinoceros—*vardhranasa* in the text, as one of the items in the banquet that Hanuman sees spread in Ravana's dining hall.[86]

It may, however, be pointed out that the interpretation of the term "*vardhranasakas*" as rhinoceros does not appear in the popular translations of the *Ramayana*.[87] For instance, the Goldman translation refrains from interpreting the term and uses it verbatim, expounding that no commentator could specify with certainty what sort of a creature it was.[88] Hence, notwithstanding the thrill of finding the rhinoceros amidst the gastronomic spread for Ravana, the presence of the animal in the text in general, and in the banquet hall of Ravana in particular, should best be considered a possibility rather than a certainty.

The *Adiparvan* of the *Mahabharata* preserves a rather quaint reference to the animal. Vashistha consecrates the Paurava as the sovereign of all baronage "to become the one horn (*vishanabhutam*) of the entire wide earth".[89] The expression lacks clarity but has been interpreted as a reference to the horn of the rhinoceros, a symbol of uniqueness and solitude.[90] It is, however, the *Karnaparvan* which unambiguously establishes the identity of the animal as our one-horned hero, when it describes a grim contest between Bhima and Ashvatthama, where the latter strikes the Pandava on the forehead with an iron arrow. Bhima is then said to have borne the arrow protruding from his forehead like a "proud rhinoceros bears his horn in the forest".[91]

The epic saga also evinces an interest in the utilization of the animal itself. Karna is consecrated according to scriptural prescriptions on a seat covered with linen, with sanctified golden and earthen pots, with water-filled tusks/horns of elephants, rhinoceroses, great bulls and other animals filled with jewels and pearls and pleasant-smelling herbs.[92] Yudhishthira questions Bhishma regarding the duration of offerings which gratify the ancestors, and is told that the gratification received from the flesh of the rhinoceros (*khadgamamsa*) was inexhaustible.[93]

The animal finds use even in the *Arthashastra* of Kautilya which spells out as forest produce: "Skin, bones, bile, tendons, eyes, teeth, horns, hooves and tails of the lizard, *seraka*, leopard, bear, dolphin, lion, tiger, elephant, buffalo, *camara*, *srmara*, rhinoceros, bison and *gavaya*, and also of other deer, beasts, birds and wild animals."[94] *Nistrimsa*, *mandalagra* and *asiyasti* are swords whose hilts are made of the horn of the rhinoceros and buffalo, the tusk of the elephant, wood and bamboo-roots.[95] It is further expounded that "a coat of mail of metal rings or metal plates, an armour of fabrics and combinations of skin, hooves and horn of dolphin, rhinoceros, *dhenuka*, elephant and bull are armours."[96]

Beyond the use of its flesh and body parts, the animal seems to have been used in contests to entertain the sovereign as well as a befitting item of gift for him. Chandragupta Maurya enjoyed seeing animal fights in his arenas. These included wild bulls, tame rams, rhinos, tusked elephants and more. We are also told that the animals brought to the king as gifts included stags, antelopes, gazelles, oryxes and rhinos.[97]

Our next encounter with the armoured giant is in the fifth pillar edict of Emperor Ashoka (268–232 BCE). Amplifying the principle of *ahimsa* or non-destruction of life, the proclamation suggests that human depredations on wildlife had begun as it decrees: "[When I had been] anointed twenty-six years, the following animals were declared by me inviolable, viz. parrots, *mainas*, ..., the rhinoceros, white doves, domestic doves, [and] all the quadrupeds which are neither useful nor edible."[98] Ashoka's word for the rhino is *palasate*.

Garbled allusions to the animal also occur in the fabulously exaggerated classical Western accounts hovering between legend and history. Ktesias (400 BCE), we know, earned himself the distinction of being the first writer to give to the Greeks the only systematic account of India till the time of the Macedonian invasion. The "horned wild ass" he elaborately describes seems to be the rhinoceros, though not all the details furnished by him correspond with the animal. Notwithstanding the muddled nature of the reference, it is the first possible allusion to the Indian rhino by a Western author.[99] A brief reference to the "one-horned Indian ass" by Aristotle (384–322 BCE), a contemporary of Alexander of Macedon, is also a likely allusion to the Indian rhinoceros.[100]

In the histories of Alexander compiled several centuries after his death, it is from Quintus Curtius Rufus (100 CE) that we learn that, following the arduous yet memorable triumph at Hydaspes, when Alexander was rousing his soldiers for the conquest of the East, they were told that the region was abundant in timber and also had the rhinoceros, an animal rarely found elsewhere.[101]

Despite its occasional inaccuracies, the *Indika* of Megasthenes, which can be retrieved only through the works of later writers, is perceived as more reliable than the mosaic of images culled from writers who never visited India themselves, and were writing on the basis of wisdom received from diverse sources. Strabo (circa 60 BCE–19 CE) for instance, cites Megasthenes when he mentions "one-horned horses with heads like those of deer".[102] The observations of Megasthenes can also be found in the work of Claudius Aelianus (Aelian), who lived around the middle of the 2nd century CE, and described a one-horned animal called the *kartazon*. The description, though inaccurate, unmistakably relates to the rhinoceros, referring to its horn and solitary behaviour, and associating it with secluded pastures.[103]

The horn of the rhinoceros, which seems to have enthralled without exception all classical writers, also figures frequently in the items of trade mentioned in *The Periplus of the Erythraean Sea* attributed to an anonymous writer of the 1st century CE. An embassy comprising a group of Western merchants trying to buy silk directly from the Chinese, instead of through Indian middlemen, were offered in exchange, ivory, rhino horn and tortoise shell—all available in India.[104]

In his *On the Peculiarities of Animals*, Aelian mentions a horn brought from India to Ptolemy the Second, which held three amphorae (about 26 gallons). He presumes it to have been from an ox "which grew a horn so prodigious", and attributes to report the breeding of one-horned horses and one-horned asses in India. He affirms that drinking cups were fashioned from the horns of these animals, adding that if a deadly poison was thrown in, the drinker would escape unharmed since the horns of the horse as well as the ass were an antidote against poison.[105] More captivating is the fragment where he recounts how the "great King of the Indians" chose a day every year for fighting between men as also between brute animals which were horned, and included "unicorn asses".[106]

What emerges from these narratives is the sense of awe, wonder and curiosity with which the Western world perceived the one-horned rhino. This, along with the fact that many of the classical writers had not seen the animal themselves, accounts for some of the fanciful descriptions we encounter. Nevertheless, not all can be attributed to imagination since some of the characteristics, such as the mention of the single horn, the habitat of the animal, its solitary behaviour as well as the reference to the tradition extolling the properties of rhino horn, leave little room for doubt that we are on apt terrain in our search for the protagonist of this story.

■ Rhino mother and calf in the elephant grass of Kaziranga. The gestational period of a rhino is 16 months. The calf stays with the mother for three to four years before it goes its own way.
Photograph: Theo Allofs/Minden Pictures/ Dinodia.

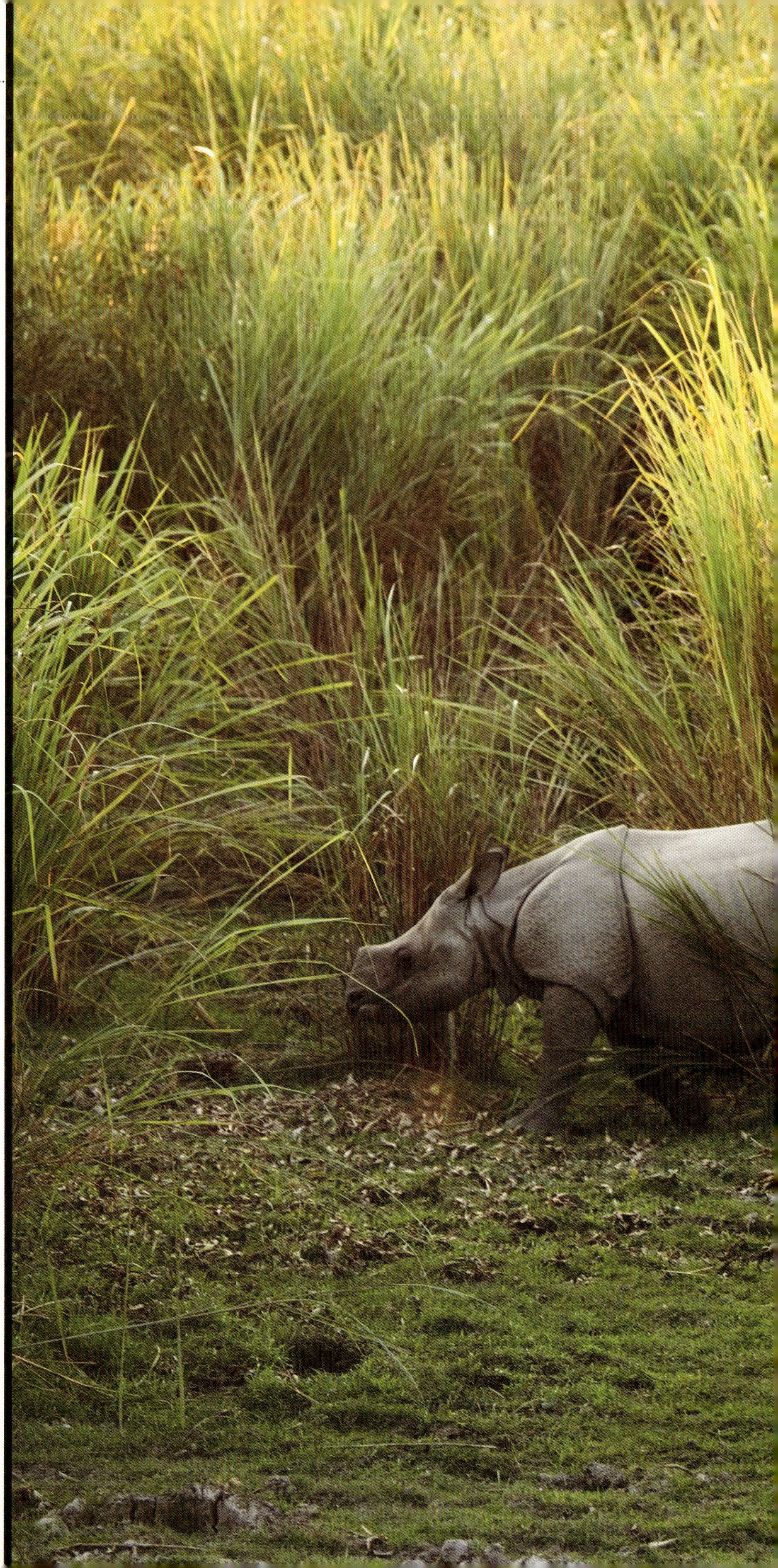

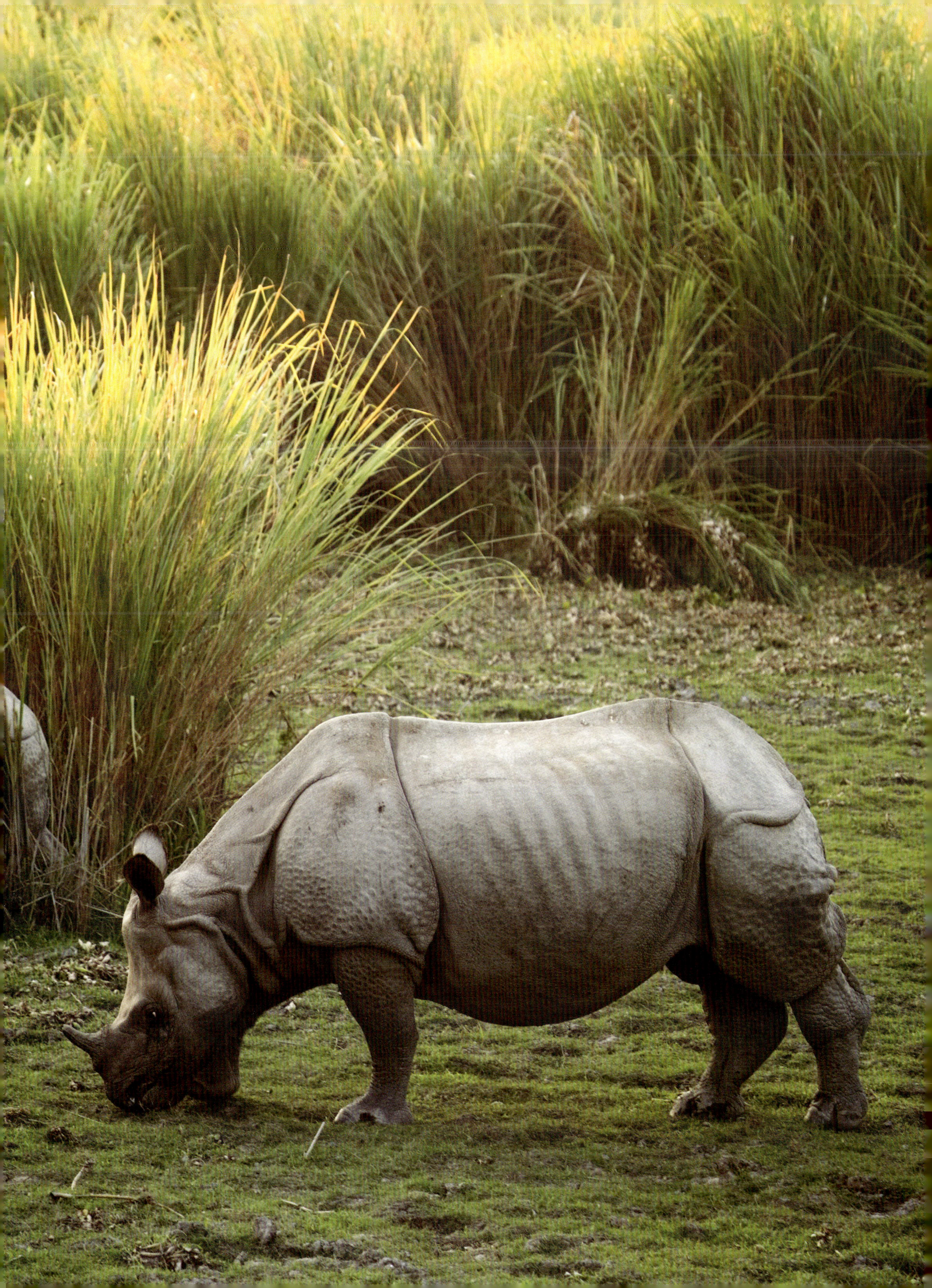

CHAPTER 4

The Unicorn and the Great Mughals

Asok Kumar Das

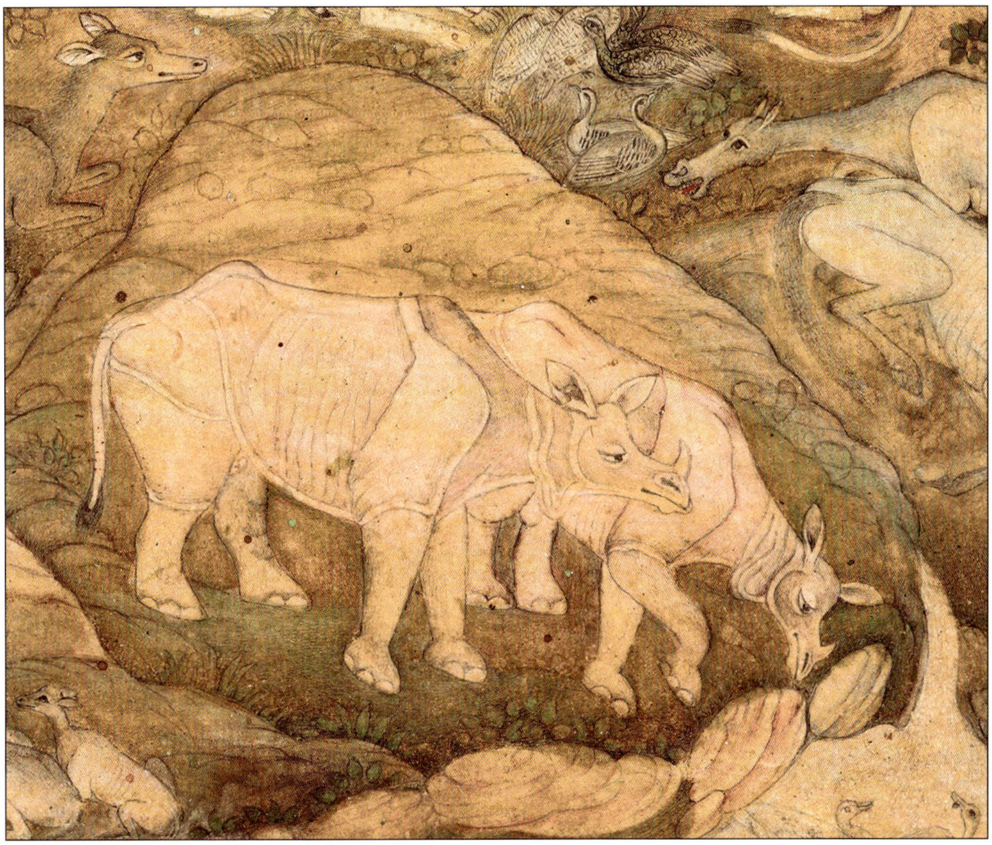

Several written and visual records of the unicorn survive from the Mughal period. The rulers of this dynasty were inveterate hunters, and Babur and Jahangir in particular were nature enthusiasts. They showed considerable interest in the rhinoceros and the animal was incorporated in several pictures along with other wild animals that were hunted in those times. However, the rhino never attained the position of the lion, tiger or elephant, possibly because the imperial hunting grounds were usually situated far away from the natural habitat of the rhinoceros, and the animal was thus met with less frequently. In this chapter we examine the representation of the rhinoceros in paintings and manuscript illustrations, most of which were prepared in the imperial ateliers under Akbar and Jahangir, the foremost patrons of these arts.

Rhinos in the *Baburnama*

Zahiruddin Muhammad Babur, the founder of the Mughal dynasty in India, being a keen observer of nature and wildlife, made a special note of the rhinoceros, an animal he had read about but not seen before coming to Hindustan. His first encounter with the rhinoceros was in the beginning of February 1519 during his entry into Hindustan at a place called Kargkhana or "rhino-home" in a thickly wooded tract close to the Swat river. He notes in his Memoirs: "Several were found, but they did not come out of the thick forest. One with a calf ventured into the open and began to run away. Many arrows were shot at it, but it managed to get itself back into the thicket. Even when fire was set to the forest, it could not be caught." An unfortunate outcome of this not-very-unusual effort was the tragic death of the calf that lay writhing on the ground. Babur's companions had to be satisfied with slaughtering this nearly dead animal to get their "trophy share"![1] This episode was chosen for illustration by the artists in three out of the four illustrated manuscripts that are known to exist of the Farsi version of Babur's Memoirs, originally written in his mother tongue Chagatai Turki. This Farsi translation with the title *Baburnama* was prepared on Akbar's orders by his principal courtier Abdur Rahim Khan-i-Khanan in 1589.[2]

If this episode was illustrated in the first copy, believed to be the "signature" copy prepared for Akbar, it remains untraced, the pages being dispersed. The second copy of the manuscript in the British Library has an illustration that shows Babur riding a white horse, watching the rhinoceros calf being butchered, while two rhinoceroses in the upper ground witness the gory sight from behind the burning forest and two more in the lower foreground flee the scene.[3] In the third copy divided between the Walters Art Museum, Baltimore, and the State Museum of Oriental Art, Moscow, the illustration shows Babur riding a fine chestnut-coloured horse and holding a falcon in his right hand, inspecting the slicing of the flesh of the young rhino, whose severed head lies in the lower foreground. As many as six full-grown rhinos huddle together behind the flames of the fire set by Babur's men in the upper background, while two more are shown leaving the scene in the lower foreground. Like the version in the British Library manuscript, this picture shows the animals and the hilly wooded landscape teeming with members of the royal party in a realistic manner. It appears that the artist, whose name is unfortunately cut out by the binder's knife, had a good idea about the physical appearance and behaviour of the rhinoceros, though he was working at least seven decades after the incident and would have had to draw the area and the people present there from imagination.

▪ Babur on a rhinoceros hunt near Bigram, attributed to Bishandas, folio from the first illustrated *Baburnama*, c. 1590, mounted on an album page of the *Farhang-i-Jahangiri* (1608–23). *Museum Rietberg Zurich, RVI 1800. Gift of Barbara and Eberhard Fischer. Photograph © Rainer Wolfsberger.*

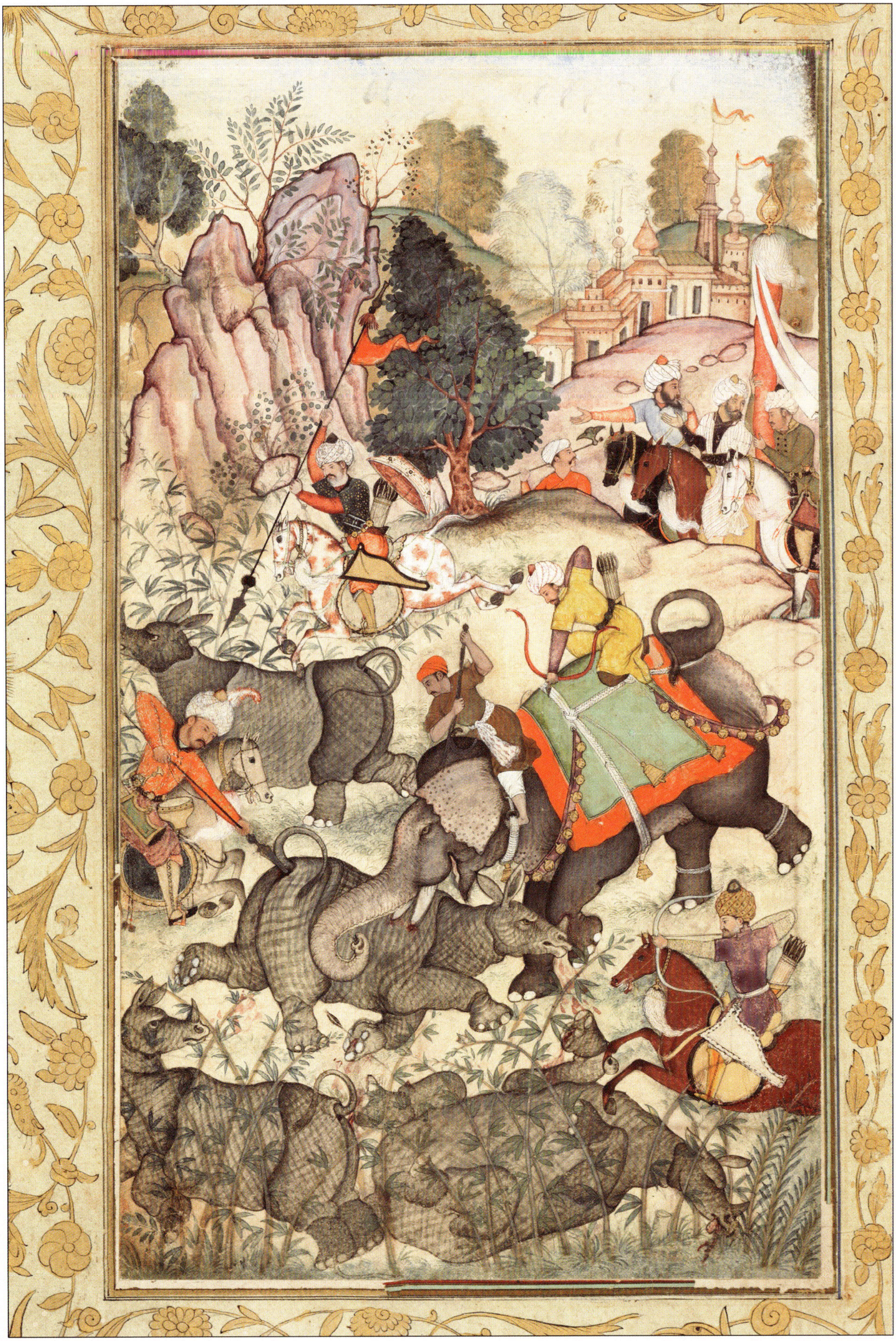

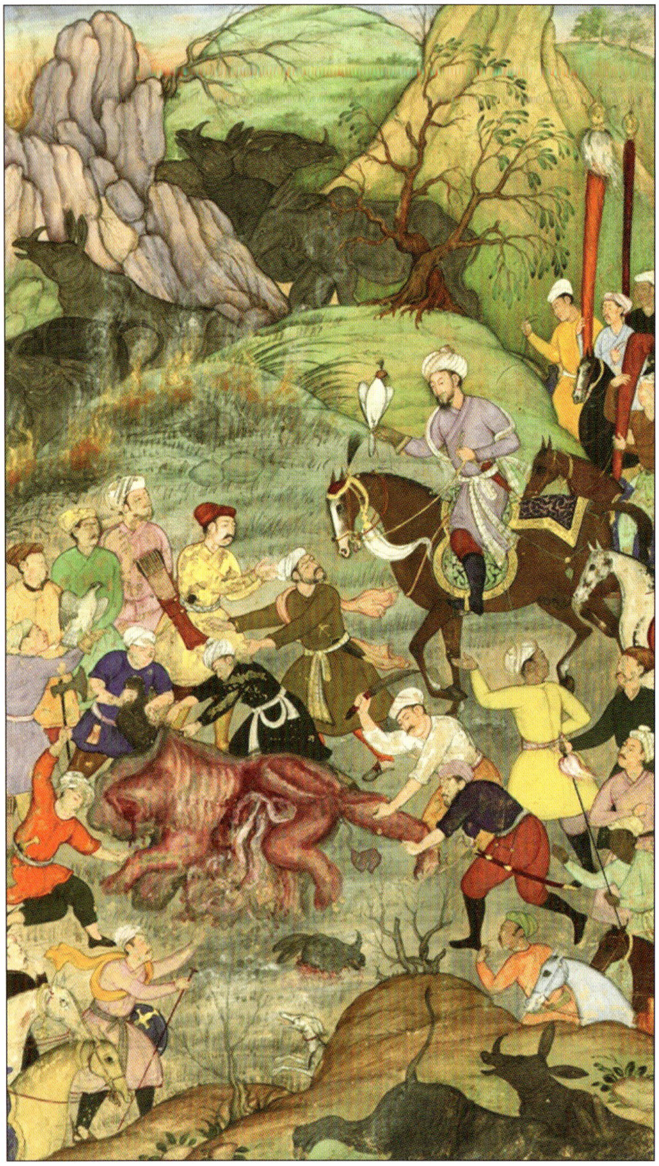

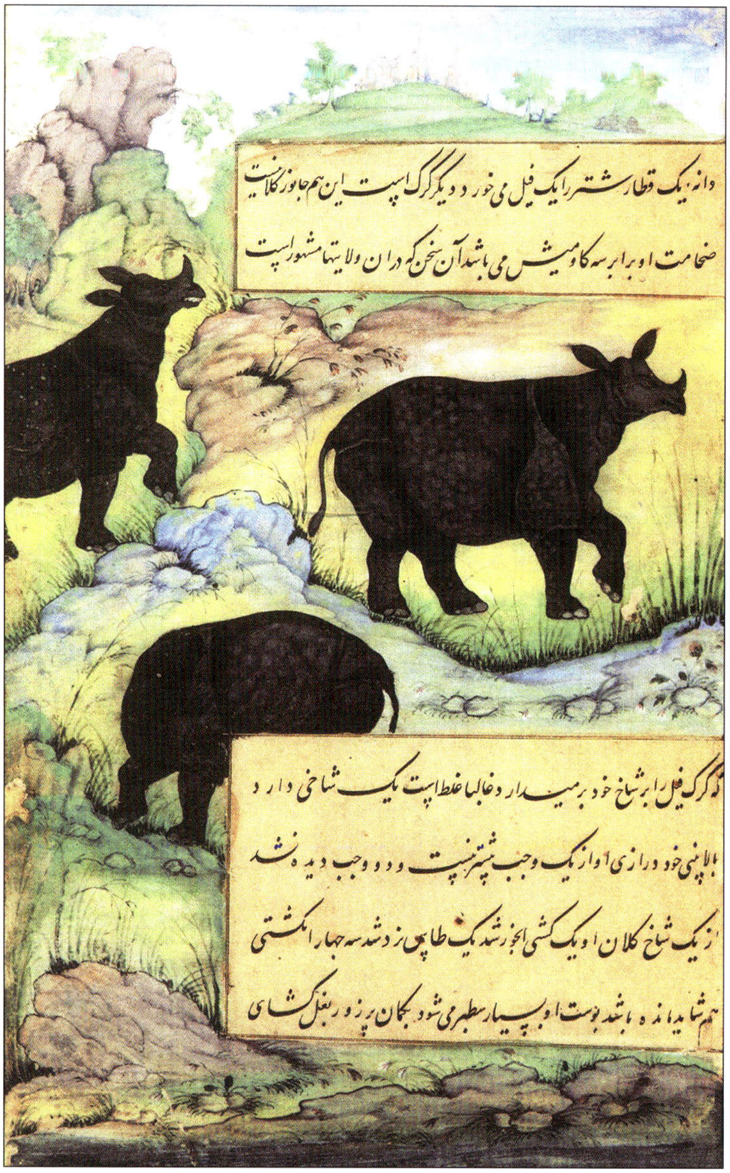

▪ Babur on a rhinoceros hunt near the river Swat, illustration from a copy of the *Baburnama*, c. 1590.
Walters Art Museum, Baltimore MD.

▪ Rhinoceroses as described by Babur in a folio from the *Baburnama*, circa 1590.
State Museum of Oriental Art, Moscow.

In the miniature drawn by the artist Kesav Kahar in the fourth copy of the *Baburnama* dated 1596, now at the National Museum, New Delhi, the details are somewhat abbreviated and the number of animals shown behind the burning bushes in the upper horizon has been reduced to two.[4]

Babur's next encounter with the rhinoceros was in the beginning of 1526 when he was camping at Bigram. According to Babur, there was no dearth of rhinoceroses in this area spreading up to the jungles of Hashnagar (near Peshawar). After crossing the Siyah-ab river, Babur was informed about the presence of a rhinoceros in a small patch of forest where it was encircled by the noisy beaters. At that time Humayun and his companions arrived from the other side of the Hindu Kush and "stood watching with delight" as they had never seen this animal before. It escaped from the ring, and was brought down after a hot pursuit for "nearly a kos", followed by two more that were also killed.

Babur was somewhat surprised to note that the rhinoceroses did not display any sign of ferociousness and avoided charging the horses and elephants driven towards them. This episode is graphically presented in all four illustrated copies of the *Baburnama*, in one single-page and three double-page pictures. The right half of the double-page illustration of the first copy is available for study at the V&A Museum.[5] In a composition occupying less than half of the text area, the young prince Humayun along with his companions watches the operation from a safe distance. The recently surfaced left half is now in the Museum Rietberg Zurich and shows Babur with his entourage attacking the rhinoceroses.[6] Babur on horseback thrusts his lance into the posterior of a full-grown rhinoceros that is also under attack from another man on a horse as well as the elephant and its two riders who use lance and arrows. One rhinoceros is lying on its back, dead, while another tries to run away. A fourth animal behind Babur's horse is under attack by one of the nobles riding a dappled horse. The entire composition finished with light washes is dominated by the rhinoceroses drawn in a lively manner. It has been attributed to Bishandas, the young nephew of the senior master Nanha, who made his debut in this manuscript.

The double-page illustration of this episode in the British Library copy, ascribed to the artists Shivnath and Jagnath,[7] also graphically presents the scene. In the left half, Babur riding a chestnut horse is shown one fallen animal by an attendant, while others shoot arrows and hurl a spear at the posterior of a second animal. The painter has taken care in composing the scene as the man with the spear is shown riding a fully caparisoned elephant; Babur writes of having included an elephant in that party as he was curious about the reaction of the rhinoceros on seeing an elephant. He had heard before that a rhinoceros could lift an elephant on its horn. In the facing page the artists have succeeded in portraying the startled expression of Prince Humayun and his companions looking at the third rhinoceros from across a rivulet. While the doublespread in the copy at Baltimore/Moscow depicts more or less the same scene, the single-page illustration of the episode in the National Museum copy shows only two rhinoceroses, one lying dead in the foreground and the other in the middle ground being pursued by hunters riding on horse and elephant. Prince Humayun along with his companions here is placed close to Babur.[8]

In the chapter of his Memoirs on "Animals That Are Peculiar to Hindustan" Babur discusses the rhinoceros just after the elephant. He gives a fairly detailed and accurate description of the animal: "The rhinoceros is also a large animal, the size of three oxen.... It has one horn in the middle of its snout...longer than a span, and none longer than two spans has been seen.... Its hide is thick. Even if you draw way back with a stiff bow and the arrow hits it right, the arrow will penetrate only four fingers deep.... Around its forelegs and hindlegs the skin is loose. From a distance it looks like it is wearing a veil. It resembles a horse more than it does any other animal. As a horse does not have a large belly, neither does the rhinoceros; as a horse has solid bones in its pastern, so does the rhinoceros; as a horse has a hoof, so does the rhinoceros. It is more rapacious than an elephant and, unlike an elephant, cannot be tamed.... They wield their horns in an amazing way. During hunts they gored a lot of men and horses.... During one hunt a young man named Maqsud had his horse thrown a spear length by one. Thereafter he was nicknamed Rhinoceros Maqsud, *Maqsud-i karg*."[9]

Babur wrote about places like Peshawar and Hashnagar to the west of the Indus having a sizeable number of rhinoceroses. In Hindustan he found them between the Indus river and Bhera in Punjab and along the banks of the Saryu river (in the Kumaun region).[10] While he was not aware of the fact that rhinoceros could

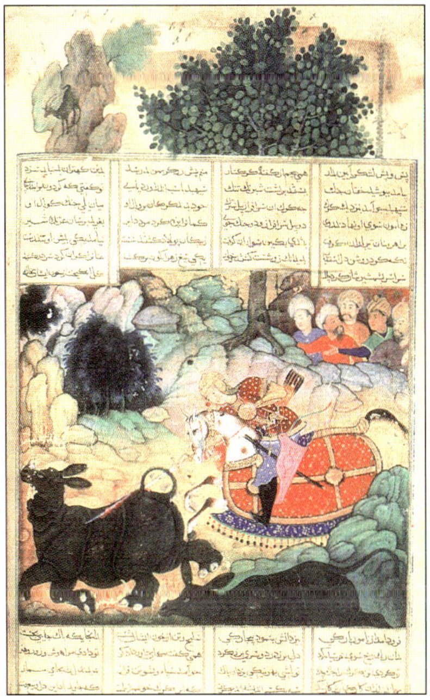

▪ Isfandiyar hunting rhinoceros, by Miskin, folio 261b from a dispersed copy of *Shahnama*, c. 1585.
Present whereabouts unknown (from Sotheby's London, October 19, 1994, lot 112).

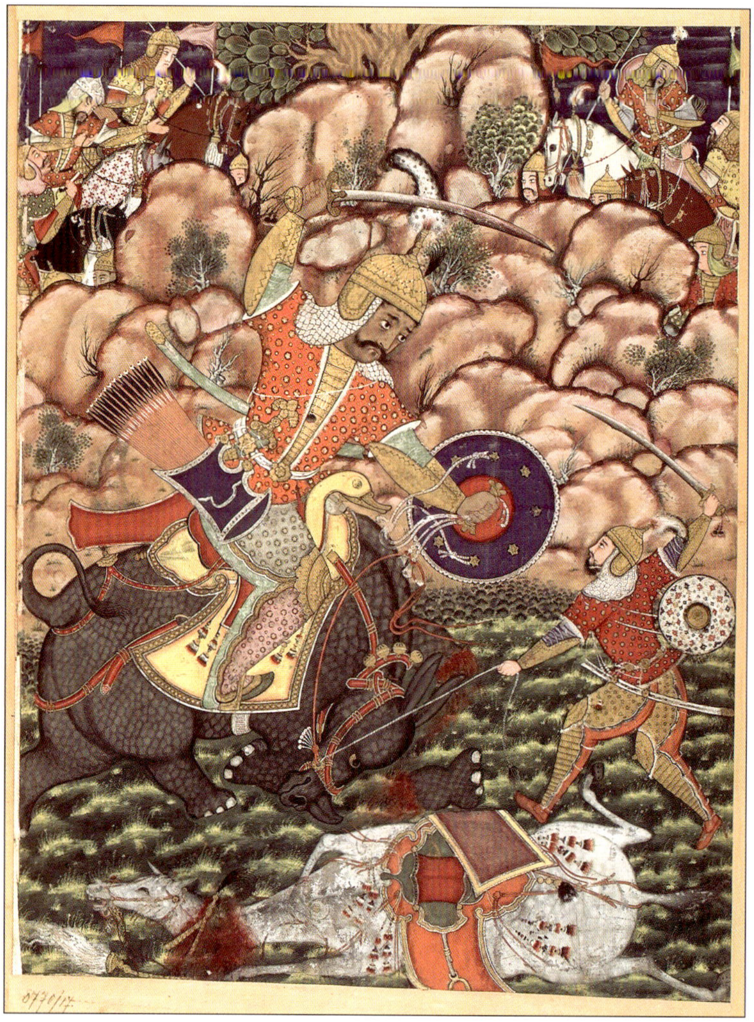

▪ A Muslim hero fights a duel against a qiqantic black warrior astride a rhinoceros, folio from the *Hamzanama*, c. 1570.
Photograph © MAK/Georg Mayer. Courtesy MAK – Austrian Museum of Applied Arts/ Contemporary Art, Vienna, B.I. 8770/17.

▪ Majnu in the wilderness, by an unknown artist, c. 1630. Single leaf, mounted on album page.
Museum Rietberg Zurich, RVI 1036, Collection Alice Boner. Photograph © Rainer Wolfsberger.

be tamed, though very rarely, he correctly observed that what he had heard in his country, that a rhinoceros could lift an elephant on its horn, was not true.[11]

This section of the *Baburnama* on the flora and fauna of Hindustan is copiously illustrated in all four known copies. The observations on the rhinoceros were picturized by the painters of Akbar's *taswirkhana* (atelier) in three of them, the National Museum copy being the exception. The illustration of the first copy remains untraced, while the picture in the British Library copy shows two rhinoceroses[12] and the Moscow copy shows three.[13] The name of the artist who made this third remarkable study is not known due to the remargination of the folio.

Other Illustrated Manuscripts

Artists of Akbar's *taswirkhana* had depicted rhinoceroses long before the illustration of the different copies of the *Baburnama* undertaken in the last decade of the 16th century. Perhaps the earliest representations of a rhinoceros in Mughal painting are in two folios of the dispersed copies of the *Hamzanama* in the Austrian Museum of Applied Arts/Contemporary Art (MAK), Vienna. One is titled "Prince Qasim beheads the rhinoceros mount of the giant Keyhur Dev in the battle for control of the Shisan Dam";[14] and the second, "A Muslim hero fights a duel against a gigantic black warrior astride a rhinoceros".[15] In the large composition of the second painting, the rhinoceros occupies most of the lower foreground. It is drawn with amazing clarity

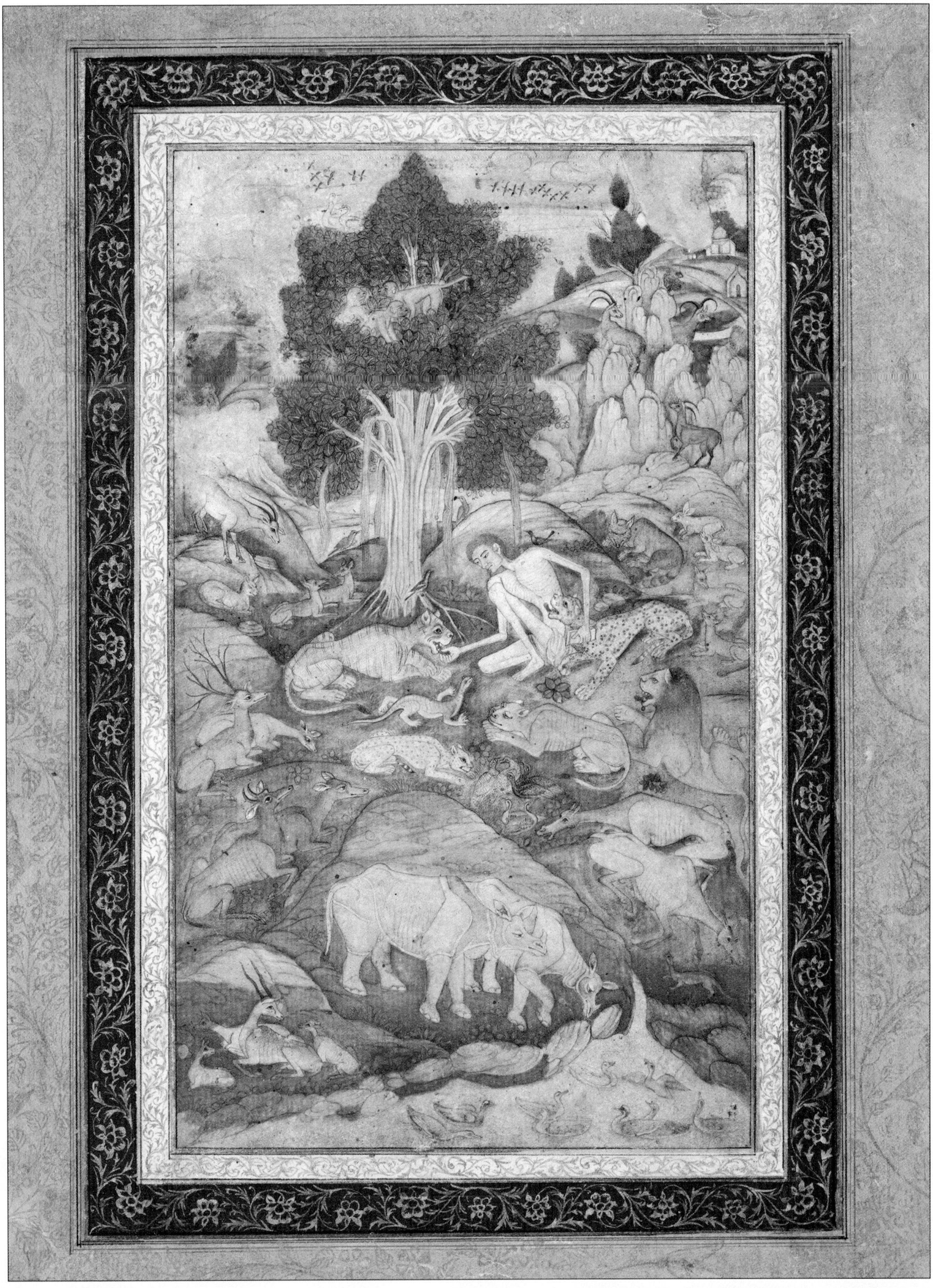

▪ Page showing a rhinoceros, from a dispersed copy of *'Aja'ib al-Makhluqat*, Delhi(?), 1789.
Freer Gallery of Art and Arthur M. Sackler Gallery, Smithsonian Institution, Washington, DC: Gift of Charles Lang Freer, no. F1907.625.

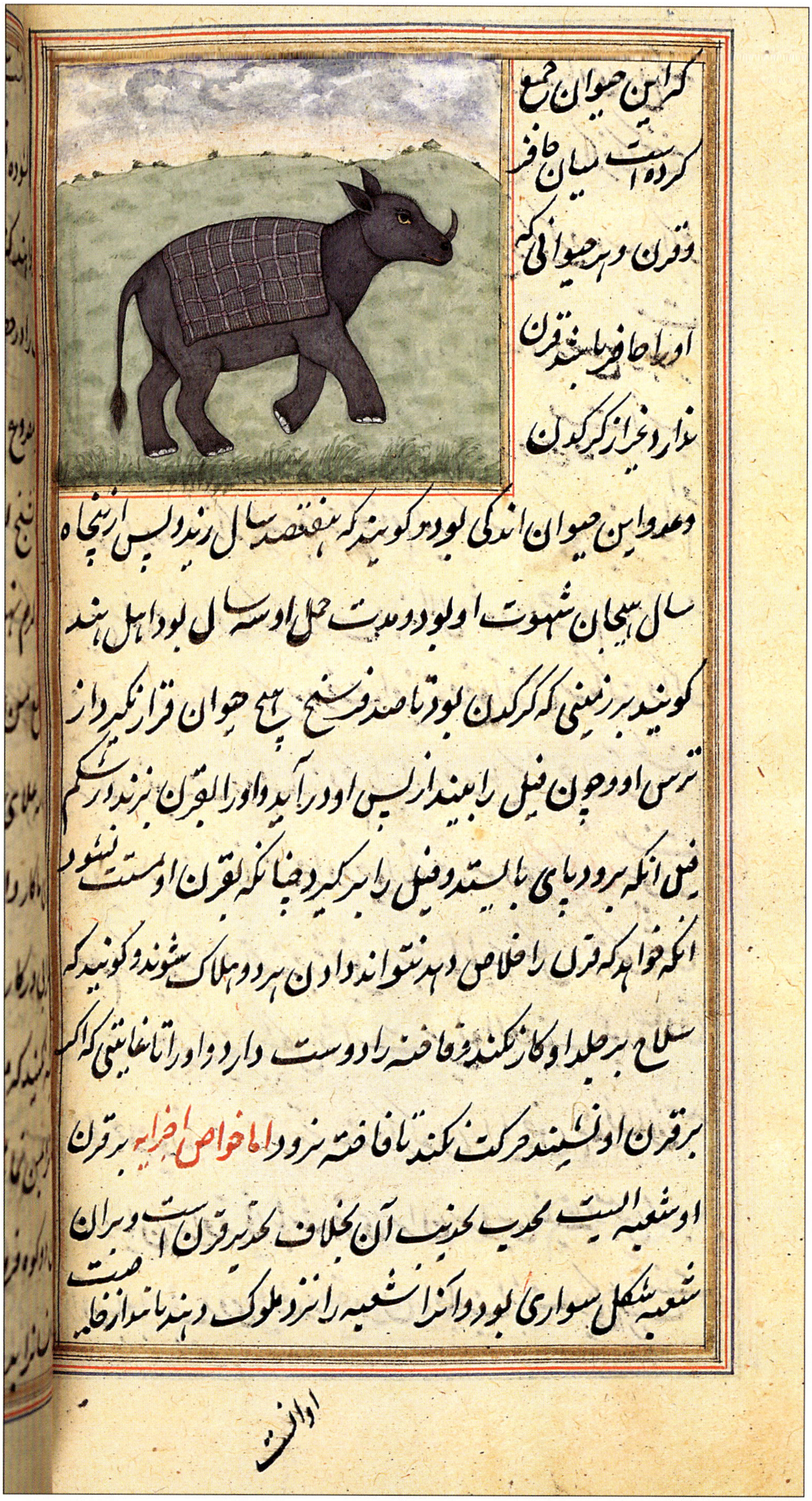

and care, with appropriate life-like colours. In spite of losing part of its left foreleg the animal has gored and killed the horse of the attacking hero.

The large-size paintings of the *Hamzanama*, executed on cloth backed with paper that contains the text, are considered as the first major artistic productions of the Mughal *taswirkhana* started under the supervision of the Persian master painters Mir Sayyid Ali and Khwaja Abd us-Samad not long after the beginning of Akbar's reign. The unnamed painters responsible for the composition and colour rendering in the two works representing the rhino appear to have been familiar with the animal but have taken some liberties to suit the demands of the work which celebrates the legendary exploits of the hero, Amir Hamza, an uncle of Prophet Muhammad, and his band of heroes. It is well-known that the hide of a rhino is extremely tough and it is almost impossible to cut through it with a sword; however one picture shows the leg amputated while the other shows the sword slicing through the neck, perhaps to demonstrate the supernatural power of the hero. Also, there is no record of using a rhinoceros as a war animal; it is not an animal that could be trained for the battlefield! The ears of the rhinoceros in both paintings look larger than life, while the markings on the skin are different in each as is the shape of the head. This is perhaps indicative of the employment of two different artists for these paintings.

In a solitary example in the manuscript of the *Razmnama*, the Farsi translation of the *Mahabharata* prepared for the personal library of Akbar in circa 1584–86, a rhinoceros is visible amongst the large number of birds, animals and plants given refuge in the giant ship to escape the Great Flood unleashed to cleanse the Universe.[16] The treatment of the animal is, however, conventional. As the artists were required to show a large number of birds and animals in small scale for this subject, they had little opportunity to depict them with accuracy. Similarly, in a painting of the episode Burning of the Khandava Forest, in another copy of the same text prepared for an unnamed patron in 1604, the unknown artist has included a rhinoceros amongst the animals fleeing the forest fire. Again here the style is less sophisticated, the treatment summarily done and the rendering devoid of life.[17]

The task of drawing rhinoceroses to illustrate the *Baburnama* manuscripts brought about greater familiarity with the animal, and the artists of Akbar's *taswirkhana* began to depict them in illustrations to literary works and individual pictures with greater confidence. The most remarkable example is to be found in the picture of Isfandiyar hunting rhinoceros from a dispersed copy of the *Shahnama* of circa 1585. This illustration is by that well-known master Miskin who excelled in drawing birds and animals.[18] It shows two full-grown rhinoceroses: one has been wounded by Isfandiyar with his spear, while the other hides beside the hilly tract. While illustrating this episode and some related ones the painters often interpreted the Farsi word *kark* or *karg* as a rhino, though in numerous other examples of the *Shahnama* prepared in Persia the animal being hunted is shown as a wolf with horns.

Instances of *Shahnama* heroes hunting rhinoceroses, or of rhinoceroses present in animal groups in other scenes, are not rare in Persian painting, as noted by Richard Ettinghausen,[19] Basil Robinson[20] and Norah Titley,[21] among others. Similar hunting scenes with rhinoceroses involving *Shahnama* heroes have been noticed in some later Mughal manuscripts. Two notable examples are in a *Shahnama* manuscript in the Chester Beatty Library, Dublin, written at the command of one Mirza Rustam (whose identity is not clear) at the city of Kangra in Himachal

Pradesh in 1695.[22] Clearly the artist has depended on some second-hand source as revealed by the stereotyped depiction of the animals.

In scenes from Nizami's *Khamsa* the rhinoceros often finds place amongst the wild animals assembled near Majnu as he wanders in the wilderness, or amongst animals in stance before Aflatun (Plato) playing enchanting music. The finest image of this genre is found in the picture of Aflatun playing music (in this case a European organ!) in the deluxe copy of the *Khamsa-i-Nizami* completed for Akbar in 1596, now in the British Library.[23] This is true of the numerous images of rhinoceroses found in compositions showing an assembly of animals, be it in the scene of Majnu wandering in the forest, or the meeting of Layla and Majnu in the wilderness, or Suleiman (Solomon) presiding over the animals, or the animal kingdom being addressed by a raven. In earlier examples drawn by masters like Miskin the composition is thoughtful and sensitive, with the animals, mostly in pairs, drawn with care and liveliness. In the Majnu scenes the animals have assembled to express sympathy for the forlorn lover; and where the lovers unite in the wilderness the animals assemble in close proximity to demonstrate their joy. The rhino is also there, with other ferocious animals, standing or squatting before them.

Amongst many examples of Layla-Majnu scenes in the wilderness drawn in various centres at various times, the ones in the Museum of Fine Arts, Boston,[24] National Museum, New Delhi,[25] and Museum Rietberg Zurich[26] stand out for their neat execution. While the first one probably belongs to the Akbari atelier the second one appears to be from a sub-imperial atelier like that of the Khan-i-Khanan, Akbar's minister who was also a poet. The pair of rhinoceroses in the Alice Boner copy of later date in the Rietberg, though lightly coloured, looks very lively, with a pair of rhinos in the lower part of the composition. Mention should be made also of two excellent late 17th- and late 18th-century pictures in the Bodleian Library, showing the union of the lovers Layla and Majnu in the wilderness before the birds and animals. In the former picture a pair of full-grown rather bloated rhinoceroses stand in a shallow pool in the foreground with elephants and ducks, fishes, turtles and an alligator; in the latter, painted in Avadh style, the landscape is greener and the rhinoceroses stand in a grassy patch as they normally do.[27] For painters of Avadh the rhinoceros was never a stranger as hordes of them lived in the neighbouring Terai grasslands close to the foothills of the Himalaya.

In illustrated copies of Zakariyya ibn-Muhammad Qazvini's *'Aja'ib al-Makhluqat* (The Wonders of Creation), prepared both in Persia and India, a lone rhinoceros is often shown standing in a landscape following the textual description. There are many examples available in different collections, both public and private: Norah Titley found as many as 18 examples in 12 different manuscripts from early 16th-century Timurid to mid-19th-century Qajar and Punjabi manuscripts in the British Library alone.[28] An image of a caparisoned rhinoceros standing in a landscape on a stray page of the *'Aja'ib al-Makhluqat* dated 1789, probably written in Delhi, shows the artist's lack of familiarity with the animal.[29] Devoid of the characteristic folded and armoured hide, though the horn and ears are of a rhinoceros, it looks more like an elephant without a trunk!

Prince Salim and the Rhino

So far, five paintings have been found that show Salim/Jahangir with rhinoceroses, either hunting with gun or in the field watching a hunt or inspecting the dead animals spread out on the ground. All of these are of high quality, as the case with paintings produced in his *taswirkhana*, and four of them show him as Prince Salim, not as he appeared after his anointment as Emperor Jahangir. The fifth, though of the same period, depicts a ruler presumed to be Jahangir, with a prince who is likely his son Khusrau.

After his estrangement from his father, Salim fled at the beginning of 1600 to Allahabad where Akbar had built a strong fort on the confluence of Ganga and Yamuna. Being a place holy to the Hindus and visited by an endless flow of devotees and pilgrims throughout the year, Allahabad had a unique ambience. The prince stayed there almost four years before returning to the fold at the time of his grandmother's death. He lived an eventful life in Allahabad, interacting with people from various strata of society as reflected by the variety of subjects chosen for translation from works popular in the region, and by the choice of pictures produced there that look very different from the types created in the imperial ateliers at Agra and Lahore. Salim was also keen to assert his independent self by assuming the title Shah or Padshah, issuing farmans, land grants and coins bearing his name with that title.[30] However not much is known of his day-to-day activities or events connected with his semi-independent existence there, beyond brief notices made by the visiting Jesuits or some isolated remarks made by contemporary historians or by himself in his Memoirs that he began writing after he became emperor. That he wasted his time in shikar, one of his most favourite pastimes, or in activities unbecoming of a Mughal crown-prince, is well known. He himself noted in the Memoirs that at the time of the death of his principal wife, mother of his eldest son Khusrau, he was away on a hunting trip. As rhinoceroses were available in large numbers in areas of Sambhal, Bahraich and Avadh, not very far from Allahabad, chances are that most of these hunting trips involved chasing rhinoceroses.

On close scrutiny, all five paintings along with a few others reveal several incongruities not noticed in the numerous hunting scenes found in historical or genealogical manuscripts like those of the *Tarikh-i-Khandan-i-Timuriyya*, *Jami-ut-Tawarikh*, *Baburnama* and *Akbarnama* produced in the Mughal *taswirkhana*. Robert Skelton has noted these differences and believes these to be pages of a lost album that Salim named *Shikarnama*, consisting of pictures of his various hunting expeditions and compiled in his personal atelier. About the painting Prince Salim hunting rhinoceros, in a New York private collection, Divyabhanusinh has pointed out that the study does not represent a single event but three separate events connected with royal hunting, the principal one being the rhinoceros hunt.[31] The top left-hand side shows five dead blackbucks, with two more being carried by hunters, while towards the middle upper right-hand side there are three dead chital with "the usual inaccurately drawn antlers". These animals are not likely to be found together in the wild. The principal episode is of Salim shooting a full-grown mother rhinoceros with a musket from the back of a female elephant. The rhino charges a male elephant driven by a nobleman, attacking the elephant with her horn. Her young calf who is close by shows signs of distress. The bottom left corner shows a collared cheetah attacking a blackbuck and a hunter approaching it with a knife to perform halal. Again, Divyabhanusinh points out that blackbuck would not be found in the habitat preferred by rhinoceros. The grass-filled ground close to a waterbody shown in the lower part of the painting is where rhinoceroses would normally be found.

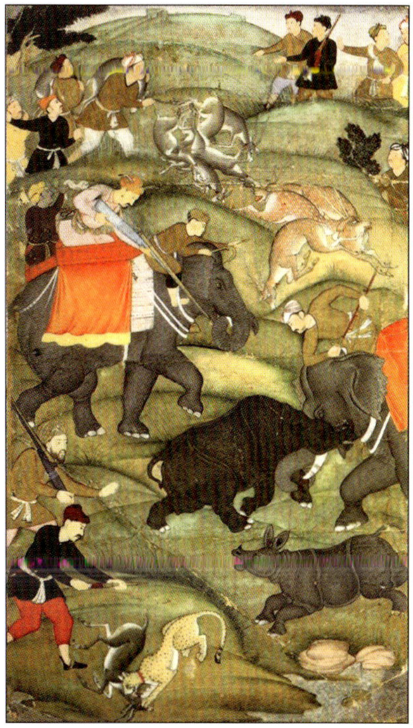

■ Prince Salim hunting rhinoceros, c. 1600–04.
Private collection, New York City.

- Prince Salim visiting the hunting field, c. 1600–04
Chester Beatty Library, Dublin, no. In 50.1.

- Prince Salim in the hunting field, c. 1600–04.
Art Gallery of New South Wales, Sydney, no. EP 1.1968.

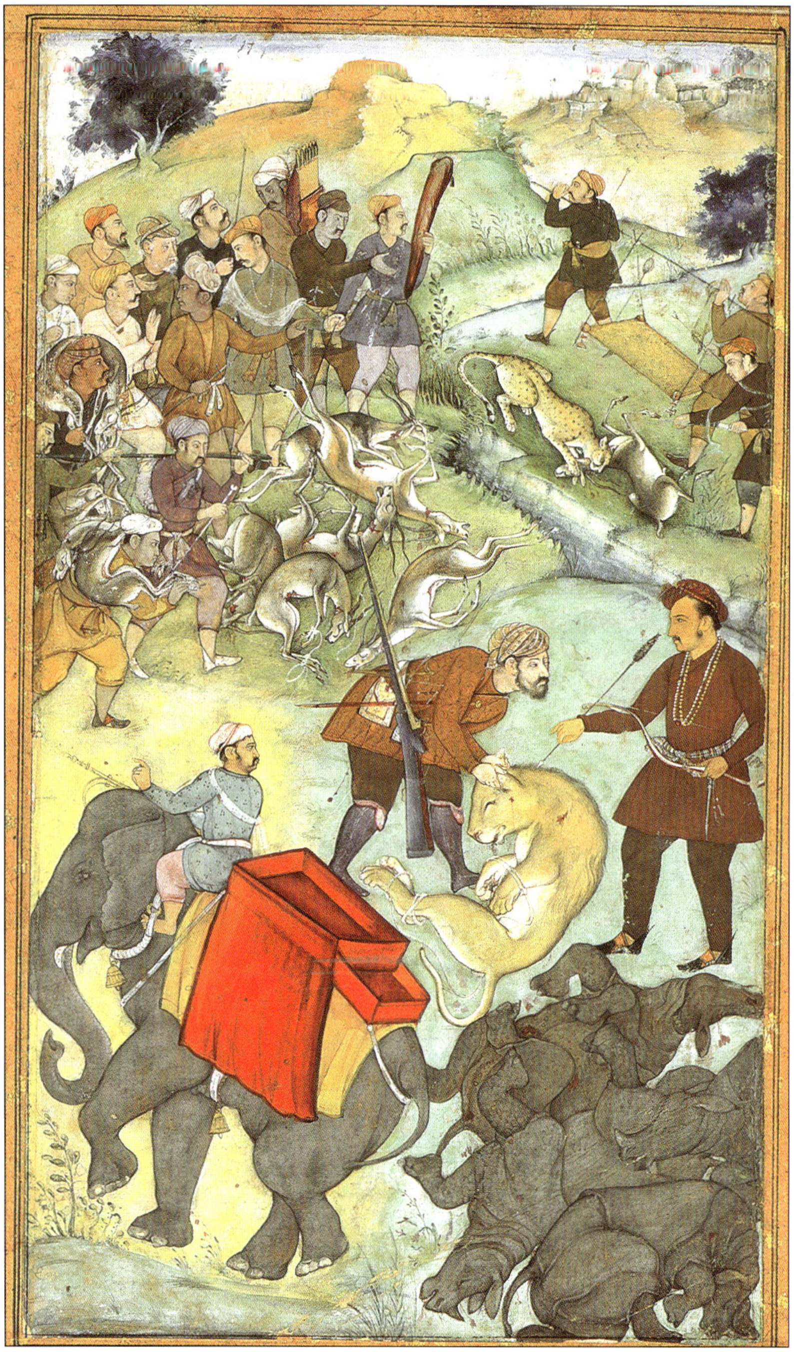

An even more incongruous combination of events has been depicted in a brilliant painting of Prince Salim visiting the hunting field, in the Chester Beatty Library, Dublin.[32] It shows Prince Salim dismounted from his elephant with bow and arrow in hand. He is talking with an old huntsman who holds the head of a dead lioness by its ears, its legs tied. The carcasses of as many as four rhinoceroses lie in a heap in the lower right corner of the composition. A female elephant with an open howdah driven by a mahout waits before them. In the upper part of the composition, on one side of a rivulet, bodies of numerous blackbucks and does lie scattered in the field before a large number of hunters, some bearing more bodies and others with arrows. On the other side of the stream a cheetah courses a blackbuck while its attendants carry its cart; there are signs of human habitation at the top right. Clearly the upper

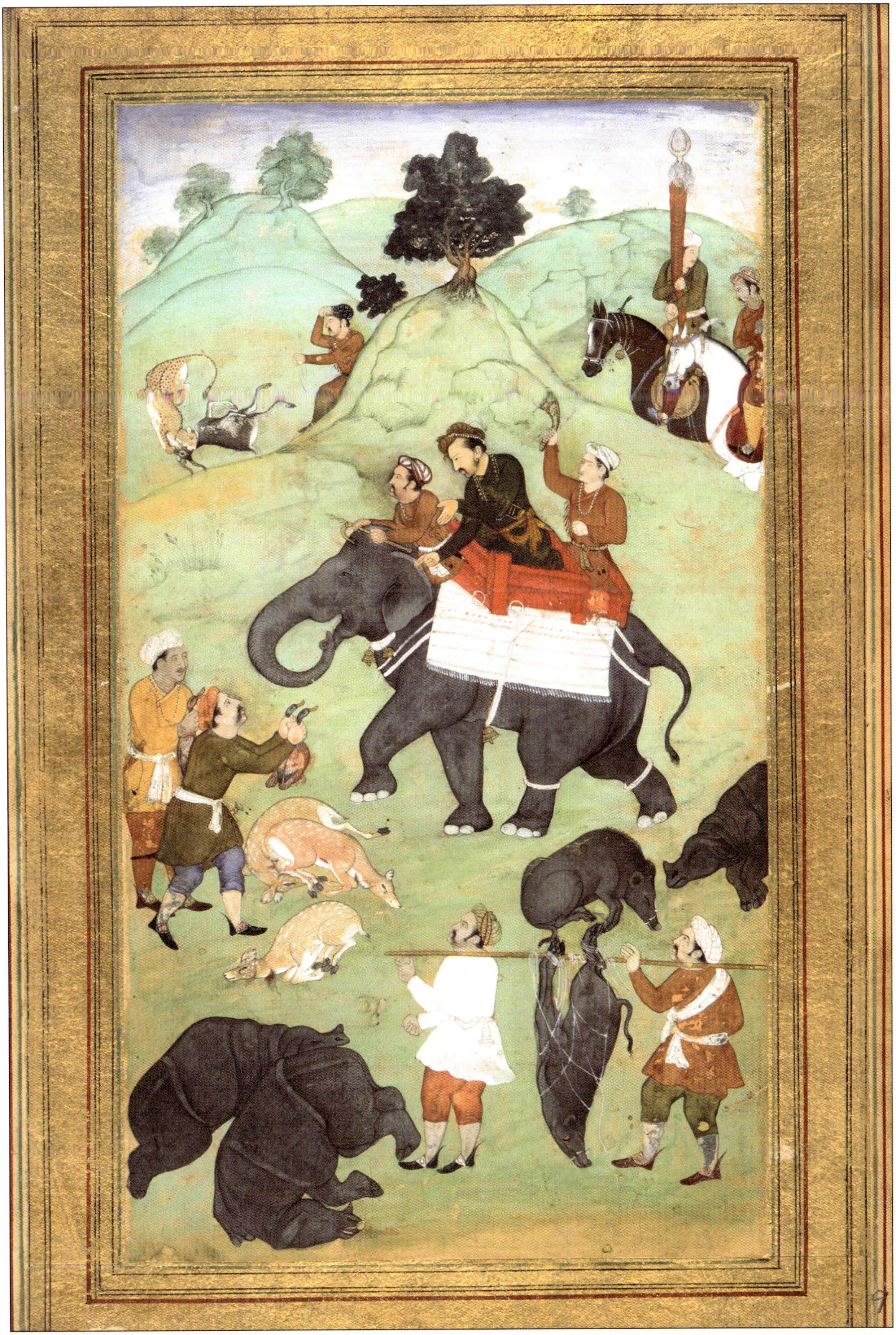

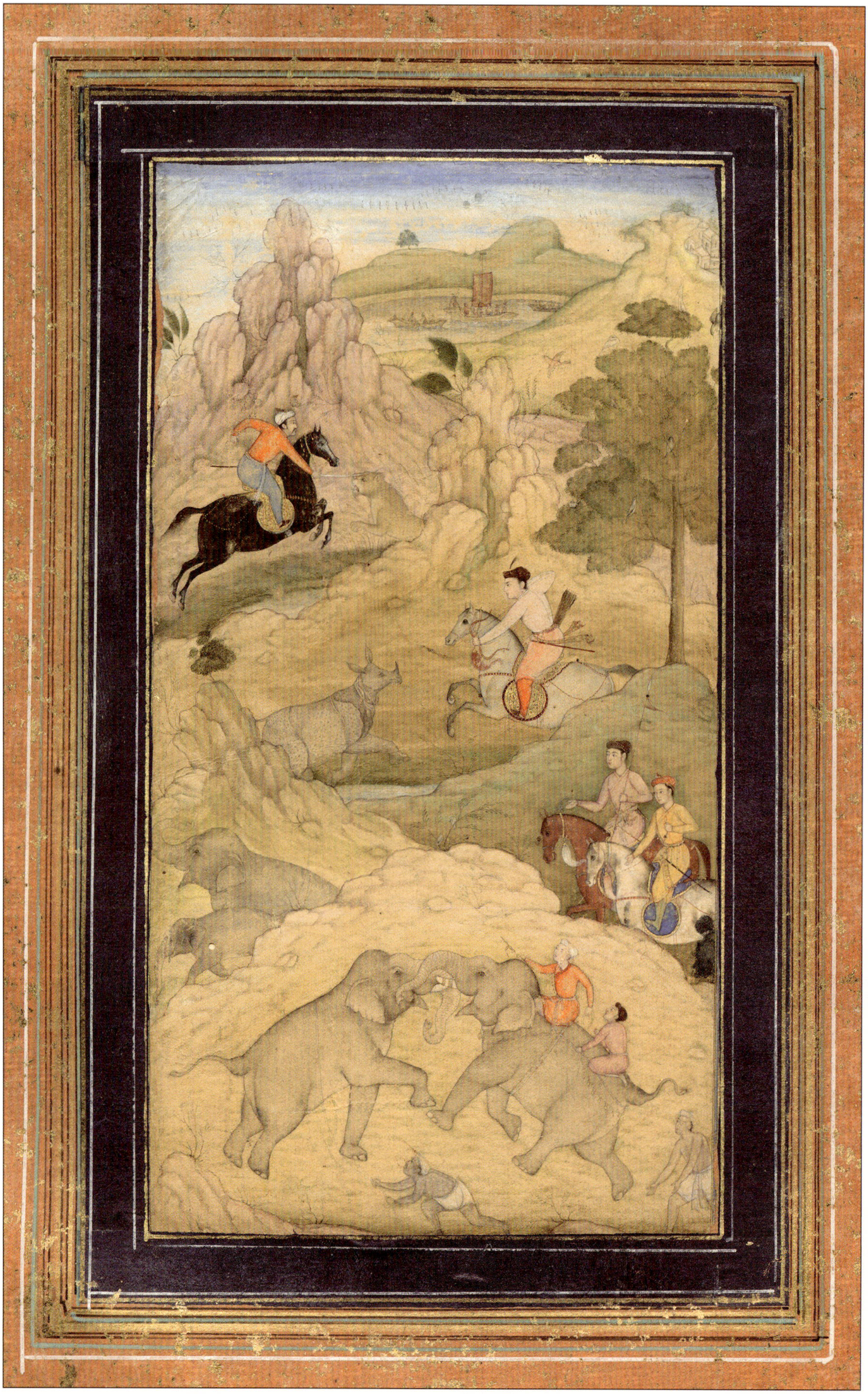

part of the painting has no relationship with the lower scene where the prince is shown, but nor are lions likely to be found in an area where rhinoceroses live. There is also no indication of how the prince managed to kill so many rhinoceroses equipped with just a bow and arrow and with no other hunter to assist him, though the old hunter bears a musket.

The third painting in this group is from the Los Angeles County Museum of Art and has close stylistic and compositional similarity with the earlier one.[33] Here Salim, seated on the back of a female elephant and holding a musket, issues instructions to the portly retainer and four others who have captured a young rhinoceros that stands still as if it is tamed. A dead rhinoceros lies in the right foreground while a dead nilgai is shown on the left. There are signs of engagement in these details of a rhinoceros hunt undertaken by the prince. Totally unrelated events occur in the upper part of the composition where two details of coursing with cheetah take place amongst the ridges of the uneven landscape in the vicinity of a city. One cheetah is running after a spotted deer followed by its keeper, while another has just pinned down a blackbuck with its keeper unsheathing the knife apparently to perform halal. As pointed out above, the events could not have taken place in the same hunting ground where rhinoceroses are found. Perhaps at some later date the painters of the *taswirkhana* may have incorporated these details above the original picture of the rhinoceros hunt to fill in the available space.

The fourth painting, in the Art Gallery of New South Wales, Sydney, also has stylistic and compositional similarity with the earlier two.[34] Here Prince Salim rides the same female elephant with the red wooden howdah fitted on a white/cream cover. He does not carry any weapon like a musket or lance or even bow and arrow essential for a hunt. He fixes his attention on a pair of mergansers caught by his retainers who stand close to three dead spotted deer before them. In the foreground two retainers carry a wild boar while another lies dead near the prince's elephant. A pair of dead rhinoceroses lie in the left bottom corner of the composition, while another lies on the right side close to the hindlegs of the elephant.

In the upper ground, partly hidden by an outcrop, are two of Salim's retainers on horseback on the right, and a cheetah coursing a blackbuck on the left with its keeper running after it. No doubt the composition is made of vignettes from different hunting scenes involving animals that do not occur in one place. As there are at least four more hunting pictures featuring Prince Salim that display such an admixture of hunting scenes, it is difficult not to agree with Robert Skelton's conjecture that an album of paintings of shikar was contemplated by the prince, which was probably left unfinished.[35]

The fifth picture of a rhino hunt was originally found mounted in an album made for Colonel John Murray who was in India from 1781. It was sold in a Sotheby's auction on June 15, 1959 and acquired by the Hagop Kevorkian Fund. It came to the art market again in 1980 (Sotheby's London, April 21, 1980, lot 133)[36] when it was acquired by Sven Gahlin, whose collection was auctioned in 2015.[37] The picture shows a prince riding a white horse hunting a full-size rhinoceros with a bow and arrow on the grassy bank of a stream. The prince cannot be properly identified though he looks like young Khusrau. This picture again contains additional hunting and animal scenes: in the upper middle ground a Mughal king who looks like Jahangir riding a dark-coloured horse is shown striking with his sword a lioness which has sprung out of a rocky outcrop to leap across the meandering stream; in

■ A prince on horseback hunting rhinoceros, attributed to Narsingh, c. 1600–04.
Private collection.

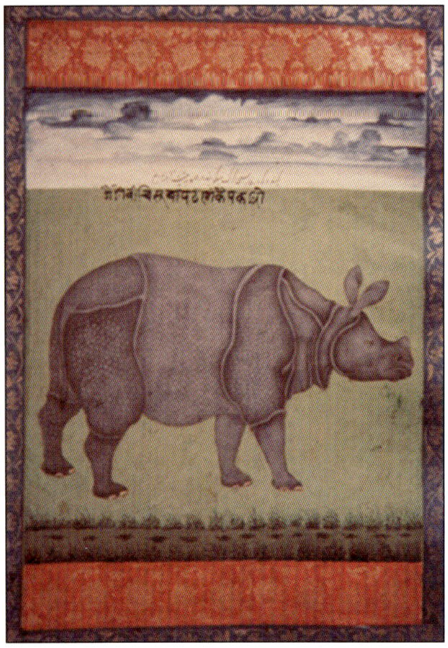

▪ Rhino brought from Suba Patna, page from the "Aurangzeb Album", c. 1665–72.
Maharaja Sawai Man Singh II Museum, Jaipur, no. AG 969.

the lower middle ground two young princes, one riding a brown and the other a white horse, follow a female elephant and her calf; and in the lower ground another young prince is shown leading an elephant of gigantic size for a fight with another big wild elephant. In the top portion of the painting is a riverscape with boats as commonly found in late 16th-/early 17th-century Mughal compositions. The composition with all these diverse scenes is very sensitively drawn with delicate colours. Such refined treatment is characteristic of the works of Narsingh who was active in the beginning of the 17th century. No doubt this one is the finest amongst the five hunting scenes discussed above.[38]

In Jahangir's Memoirs, the *Tuzuk-i-Jahangiri* or *Jahangirnama*, the emperor provides a list of animals hunted by him, which gives the number of rhinoceroses killed as 65. However there is only one reference to a rhinoceros hunt. This occurred late in his reign, in the summer of 1624, when he hunted rhinoceros near the lake in Nuh-ban identified as in Koil, modern Aligarh, not very far from Agra.[39] No visual record of this hunt has been found; this was the time when the emperor was not keeping good health and no longer showed interest in having pictures of important events made by his favourite painters. While the Memoirs contain details of many hunting expeditions as well as illustrations of these by the court artists, there is no record of any rhinoceros hunt during previous years of his reign. A possible reason may lie in the fact that during these years he moved mainly between Agra and Lahore and from there to Kabul and in later years to Kashmir. A considerable part of his reign was spent in travelling to Ajmer and from there to Gujarat and Malwa, areas where rhinoceroses would not be found. In the course of the eight decades since Babur had noticed and hunted rhinoceroses on the western bank of the Indus and recorded their presence as far as Bhera in Punjab, the rhinoceros population seems to have disappeared. Even during the brief period Babur ruled India between 1526 and 1530 there is no mention of any encounter with the rhinoceros. From what is reported by Babur in the section of his Memoirs on the flora and fauna of India, or noted by Abu'l Fazl in the *A'in-i-Akbari*[40] the westernmost limit of the rhinoceros habitat during this period did not extend beyond the eastern bank of the Yamuna.

Examples of rhinoceros depictions in manuscripts and albums produced in Jahangir's time include two folios of a rare and valuable Persian manuscript of *Yusuf wa Zulaykha* written in 1557 by the celebrated calligrapher Mahmud b. Ishaq al-Shihabi. The manuscript came to the Mughal court where it was refurbished in Jahangir's kitabkhana by decorating the margins (hashiya) with golden designs. These two folios show a pair of rhinoceroses, in the top margin of one folio[41] and in the lower margin of the second.[42] The designs on the rhinoceros hide in these two instances differ, indicating that they were not the work of the same artist.

It is known that there were rhinoceroses in the emperor's menagerie, as noticed by the English traveller Thomas Coryat in 1615–16 when Jahangir was holding court at Ajmer. The first Englishman who visited India to see the country, ride an elephant and to write an account of his travels in a book, Coryat saw a pair of rhinoceroses among various other wild animals in Emperor Jahangir's large menagerie. Coryat came to know that the rhinoceros had been brought from Bengal:[43] "Hee keepeth abundance of wild beasts, and that of divers sorts, as lyons, elephants, leopards, beares, antlops, *unicorns*: whereof two I have seen at his court, the strangest beasts of the world. They were brought hither out of the countrie of Bengala, which is a most singular fertilitie within the compasse of his dominion, about four monthes journcy from this...."[44]

The only other reference to a rhinoceros present at the Mughal court, some 60 years later, is furnished by the picture of a large male rhinoceros in an album compiled at Aurangzeb's court, now in the Maharaja Sawai Man Singh II Museum, Jaipur. The painting carries a legend in Farsi and Nagari stating that the animal was captured from Patna *suba* (district), an area known for the presence of the rhinoceros until the early decades of the 20th century. The album contains a number of pictures of rare and interesting birds and animals brought from different parts of the country and outside. Some of them are met with for the first time in this country, like the ostrich and cassowary, that made their way to the Mughal court through trade channels.[45] The precise date when the rhinoceros was brought from Patna has not been recorded, but from the dates mentioned in some of the other images the album can be dated to 1665–72. Being a rarity, the rhinoceros deserved a detailed study worthy to be placed in the album.

A few later representations of the rhino are worth mentioning here. In the Museum of Islamic Art, Berlin, there is a drawing with touches of colour which shows a standing rhinoceros with other animals like wild buffalo, nilgai, blackbuck, coursing cheetah etc. interspersed with vegetation and arranged in horizontal rows. The drawing is mounted on an album folio dated to the beginning of the 18th century.[46] Some years ago a well-drawn picture of a full-grown Indian rhinoceros was exhibited in a Paris gallery, that was attributed to one of the Deccan schools of painting.[47] It has now been found to be a 19th-century copy by a local artist from Delhi or Avadh, of an engraving of a rhinoceros that was made for Dr James Douglas by his artist-draughtsman Dr James Parsons in London in 1743.

It must be mentioned in conclusion that there was never any attempt to train the rhinoceros or breed it in captivity at the Mughal court, though, as we have seen, it was occasionally part of the menagerie. The Mughals admired shields made of sturdy rhinoceros hide[48] and used thumb rings made of rhino horn. However, they never used the rhino horn for making luxury items as the Chinese did. Jahangir showed considerable interest in procuring narwhal tooth and walrus ivory from distant lands and getting archers' rings[49] or knife-handles made from them, but he never seemed to have tried the rhino horn for similar use, though this would have been more easily available to him.

CHAPTER 5

At Large in Art and Culture

Asok Kumar Das

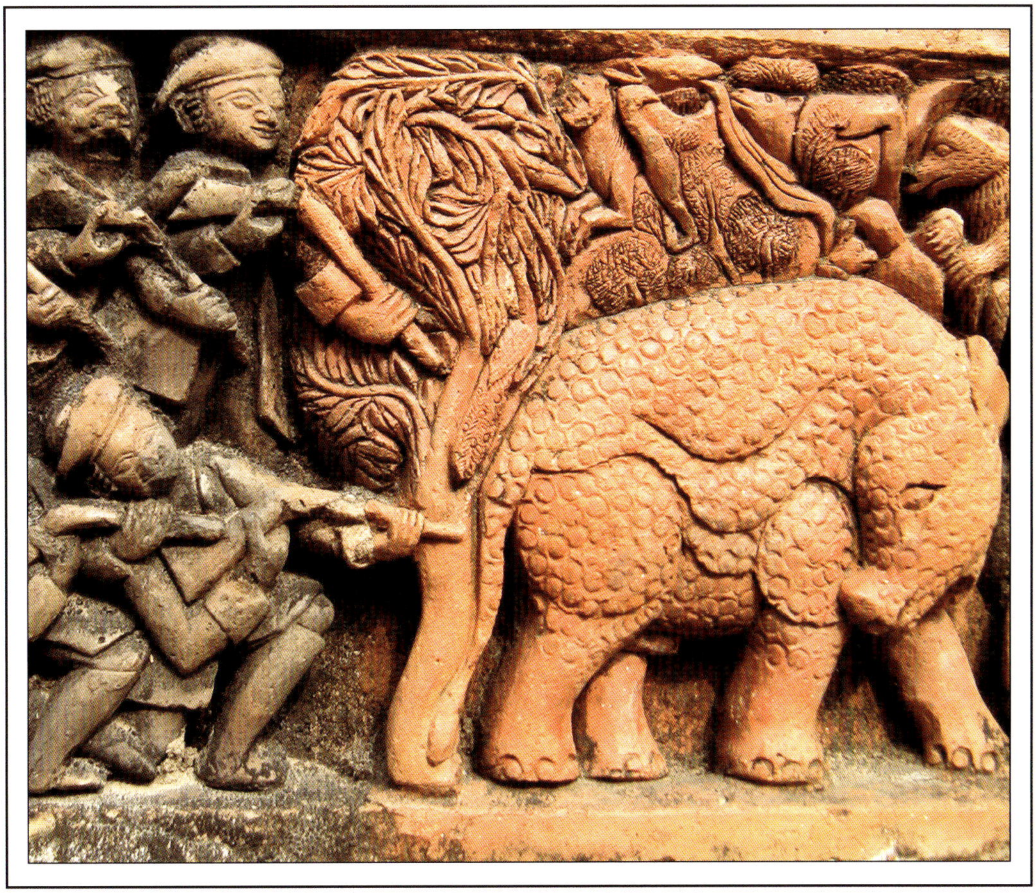

Nature with its bounty and infinite variation is present in some form or the other in most artistic creations through the ages and across cultures. This is particularly true of Indian art, where gods and goddesses, kings and queens, heroes and heroines are shown in groves and forests full of trees and creepers laden with flowers and fruits, with birds seated on their branches, and pet animals and wild beasts moving around. These living things are all depicted with great care and attention to detail. Among the wild beasts, the elephant, wild buffalo, wild boar, lion and tiger attracted the most attention as they were associated with gods and goddesses from early times. This is not true of the rhinoceros as it never enjoyed an association with any major divinity, with the exception of a relief at Angkor Wat in far-off Cambodia, where the Vedic god Agni is shown riding a chariot drawn by a rhinoceros. The probable reason is the relative unfamiliarity of the animal that is found only in wet grasslands. Besides, the rhinoceros was never domesticated or trained as a war animal.

In this chapter we survey a cross-section of depictions of the rhino, spanning paintings in manuscripts and albums as well as on palace walls, relief sculptures in stone and terracotta, carpets and textiles. Objects made from rhinoceros horn or hide and used as decorative, ritual as well as defensive gear are also noted. These works and objects represent periods not covered in the other chapters of this book.

Manuscripts, Albums and Wall-Paintings

A reasonably recognizable image of a rhinoceros is first noticed within a painted roundel on a wooden manuscript cover or patali in the Jaisalmer Jain library or Gyan Bhandar, datable in the 13th century.[1] The rhinoceros being associated with Shreyansnath, the eleventh Jina, the appearance of the animal in Jain art is not unexpected. However painted examples are rare. The only other example that can be cited here is in a folio of the *Mahapurana* in the Sri Digambara Jain Atisaya Kshetra, Jaipur, written at Palam, close to Delhi, in 1540.[2] Here the rhinoceros is shown without any religious association, roaming among trees in a forest along with many other animals.

The earliest depiction of the rhinoceros in a Persian work is in a copy of the encyclopaedic *Miftah al-Fuzala* composed by Muhammad ibn Daud Shadiyabadi for his patron Ghiyas ud-Din Khalji of Malwa at Mandu (known as Shadiabad) towards the end of the 15th century.[3] As in other early encyclopaedic works like Qazvini's *'Aja'ib al-Makhluqat* and similar texts like Muhammad al-Mufti Balkhi's *Majma al-Ghara'ib*, the rhinoceros is shown as an example of rare and strange animals and birds found in different parts of the world then known to their authors. Though the illustrations of these texts often followed earlier versions based on hearsay, the Mandu image looks different from the original.

The earliest dated image of a rhinoceros in a Sanskrit manuscript is a small illustration of a hunting scene in a manuscript of the third book, *Aranyakaparvan*, of the *Mahabharata* in the Asiatic Society, Mumbai. This incomplete copy has a colophon with a date corresponding to 1516 CE.[4] The paintings are drawn in a summary manner without much attention to detail. The chosen example shows an animal with a large head, long ears and a prominent horn. Its body is pierced with arrows as it is trying to escape from a hunter identified by an inscription as a Pandava. Though the rhinoceros is not specifically mentioned in the text, the painter no doubt was familiar with the animal as he has chosen to include it with other sacrificial forest game.

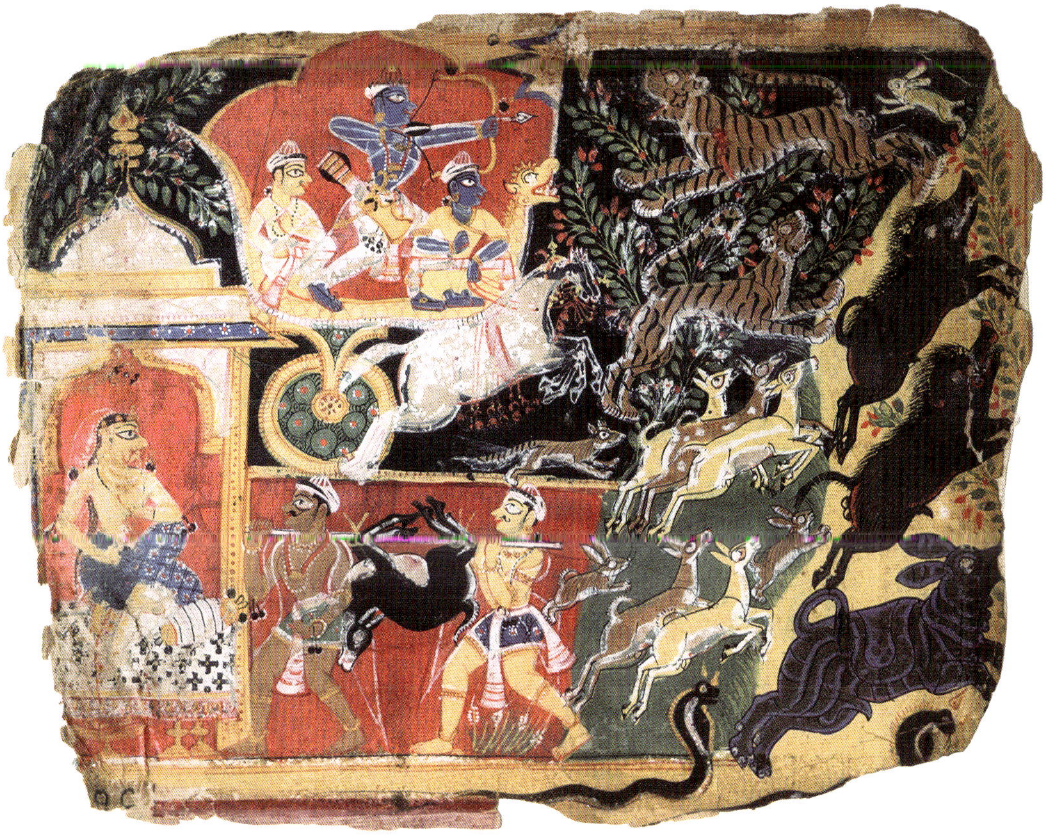

■ Krishna hunts animals suitable for sacrifice, from the Palam-Isarda *Bhagavata Purana*, mid 16th century. *Formerly in the Ehrenfeld Collection.*

Some three decades later, in a remarkable illustration of the *Bhagavata Purana* painted in a well-developed style with opulent colours for an unknown patron, the rhinoceros is depicted in the bottom right corner of a scene showing Lord Krishna in a chariot shooting arrows at wild animals that include tigers, wild boars, deer, hares, snakes, foxes and a large rhinoceros. While the treatment of most other animals follows the prevailing somewhat summary norm, the rhinoceros has been depicted with care.[5]

No image of the rhinoceros outside the imperial Mughal atelier in the following decades has been found. However, from the second half of the 17th century the rhinoceros appears in various media in different parts of Rajasthan, especially at Bundi, Kota and Udaipur.

Milo Beach published a detail from the extensive wall-paintings at Bundi's Chattar Mahal which depicts the Kota ruler Ram Singh (d. 1707) hunting a rhino. That scene conformed to a damaged but powerful tinted drawing of a similar hunting scene in the Cary Welch collection, and another high-quality drawing in a private collection, reproduced here. The drawing does not show the ruler, but shows a young prince or a mahout attacking the rhinoceros from elephantback.[6] The Kota rhinoceros hunting scene, both on the palace wall as well as on paper, shows the hunter using a spear while the elephant on which he rides traps the prey with his trunk.

Joachim Bautze made a special survey of the surviving wall-paintings in the different palaces of Kota and Bundi, and other prominent monuments in these cities as well as in Bijolia, Samod and Jaipur, and noticed the presence of rhinoceroses in profusion.[7] The Hadoti rulers of Bundi and Kota were fond of hunting in the rugged countryside, hilly tracts and forests of their kingdoms, abounding in big game, and

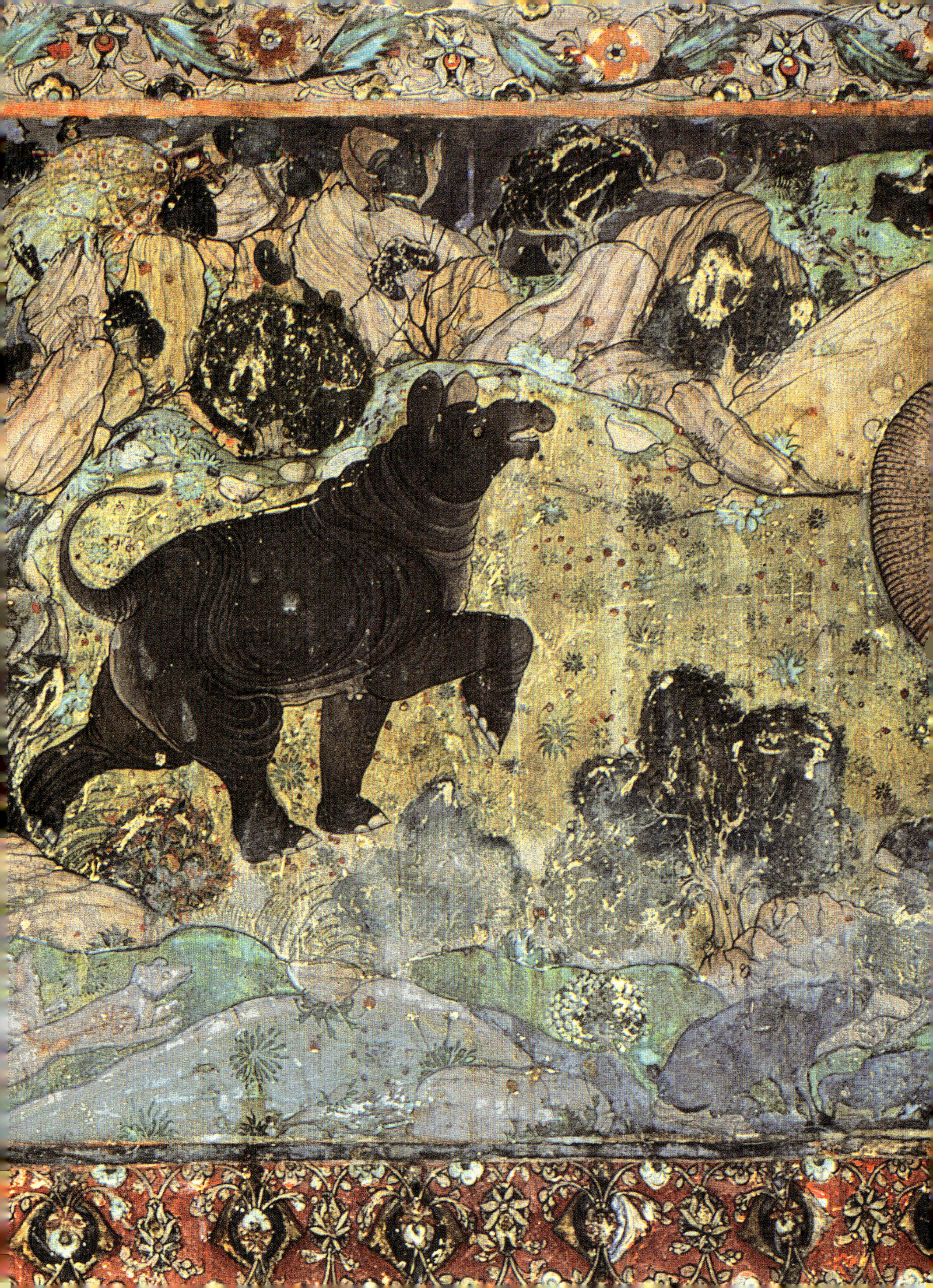

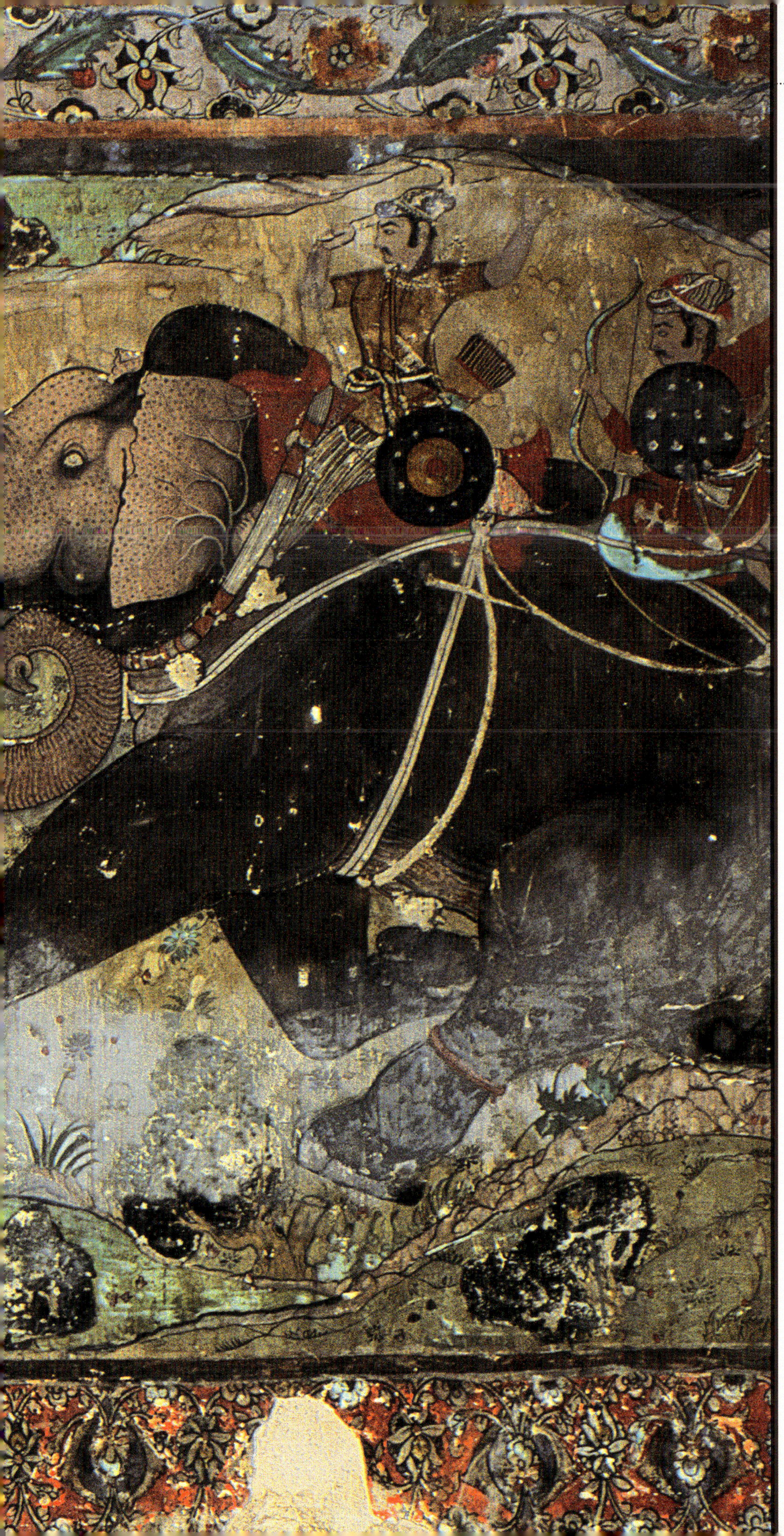

■ Bundi ruler hunting a rhinoceros, detail of wall-painting at the Bundi Palace.

Courtesy Shubha and Prahlad Bubbar.

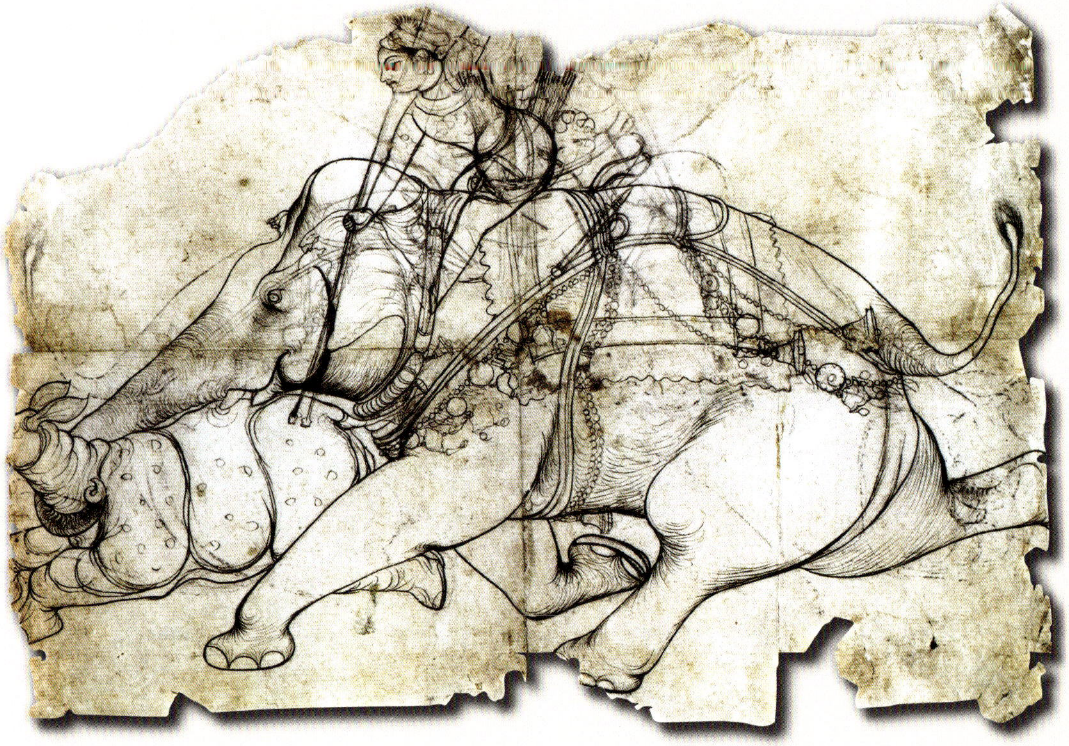

■ Prince pursuing a rhinoceros, drawing based on a painting in the Stuart Cary Welch collection, c. 1750.
Private collection.

■ Nawab hunting rhinoceros near Bahraich, page from the Gentil Album, Faizabad, 1769.
Victoria and Albert Museum, IS 25-1980.13.

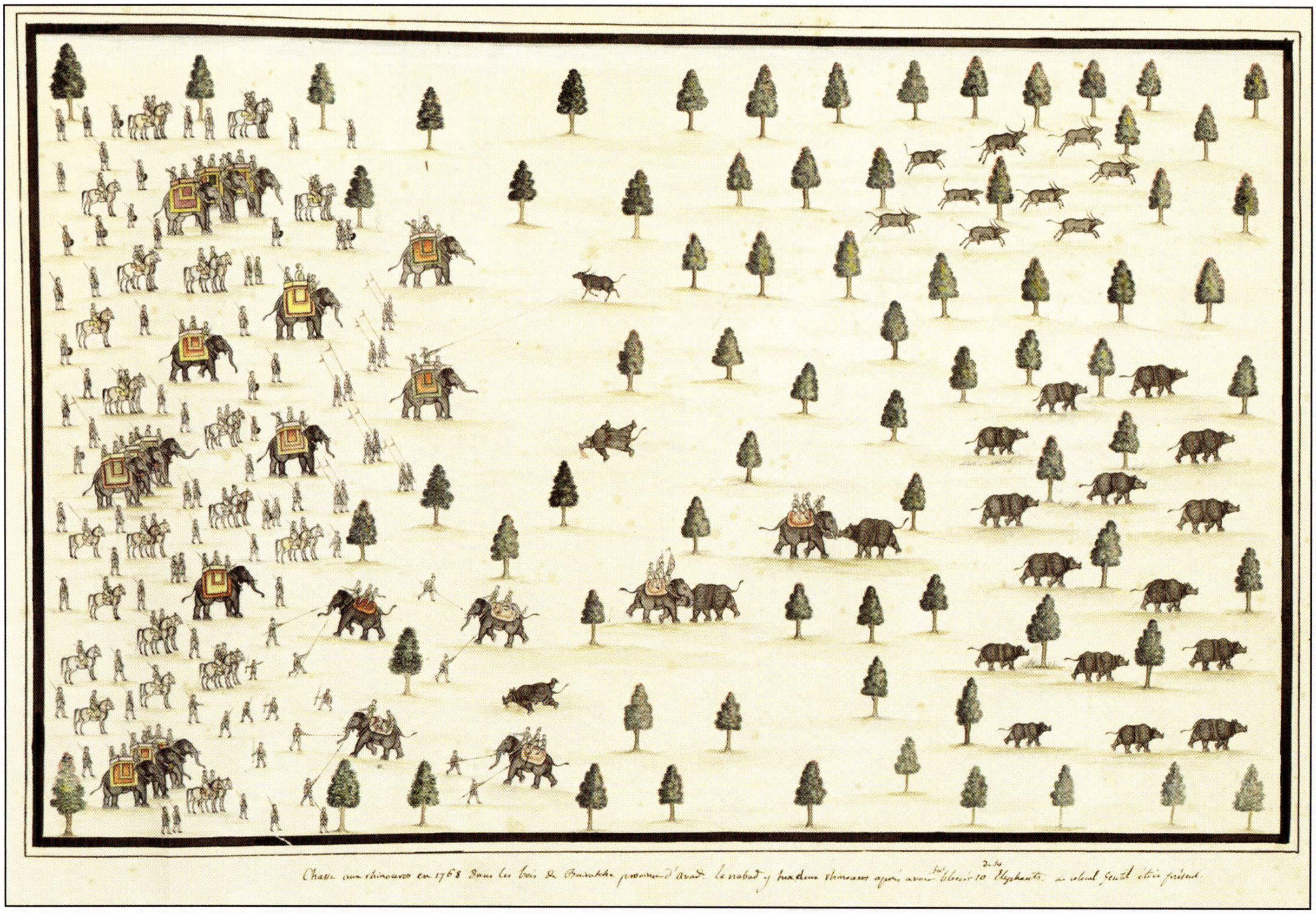

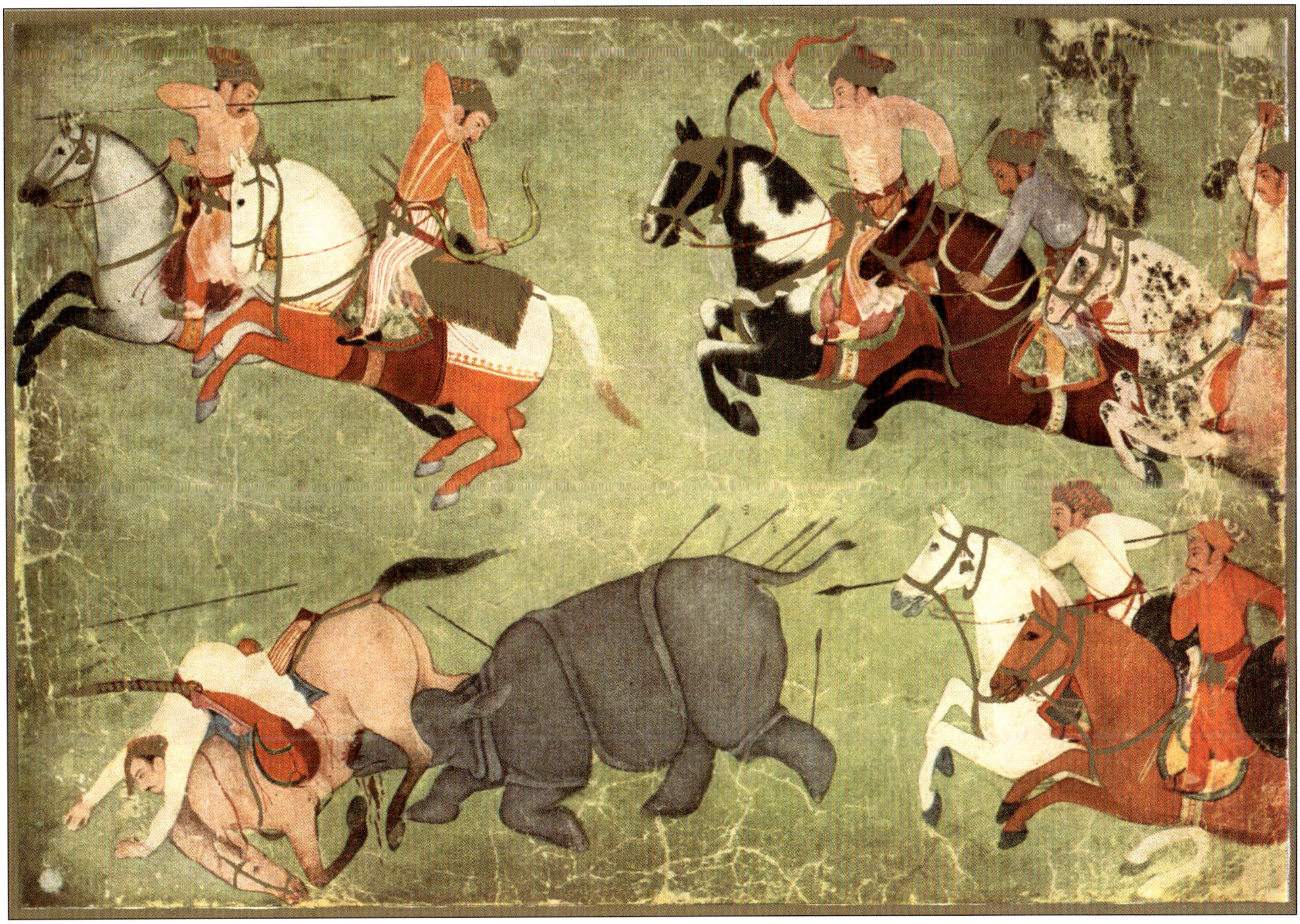

▪ Rhinoceros hunt, probably Kota/Bundi, c. 1650–75.
Sir Cowasji Jehangir Collection.

numerous pictures of these hunting expeditions were commissioned to decorate the walls of the palaces that were built by successive rulers and leading courtiers. The appearance of rhinoceros in these hunting scenes, one of which is illustrated here, is rather intriguing as no firm record of the presence of the animal in the forests and scrublands of the region has been found for this period. In fact the only record of the presence of a rhinoceros on the right banks of the rivers Ganga and Yamuna is the entry of the year 1624 in Jahangir's Memoirs, mentioned in the previous chapter, relating to his rhinoceros hunt at Nuh-ban, near Aligarh.[8] However, the Rajput rulers of Amber, Bundi, Kota and Bikaner often joined the Mughal emperors in their frequent hunting expeditions in different parts of the country, and so would have been familiar with the rhinoceros and the precise method of hunting it with spears or bow and arrows aimed at the weaker spots in the shoulder, under-belly or posterior.

There are at least four Rajput paintings on paper, one drawing and one large painting on cloth that show the rhinoceros. The first painting, from the Sir Cowasji Jehangir Collection, Mumbai, shows a group of Rajput princes riding on spirited horses cornering a full-grown rhinoceros in an open field. The rhinoceros, in spite of being pierced by arrows, has gored a horse whose rider has been thrown off. One of the hunters in the lower right riding a white horse is pointing his spear at the posterior of the rhinoceros. No doubt this is the finest painting of a rhinoceros hunt

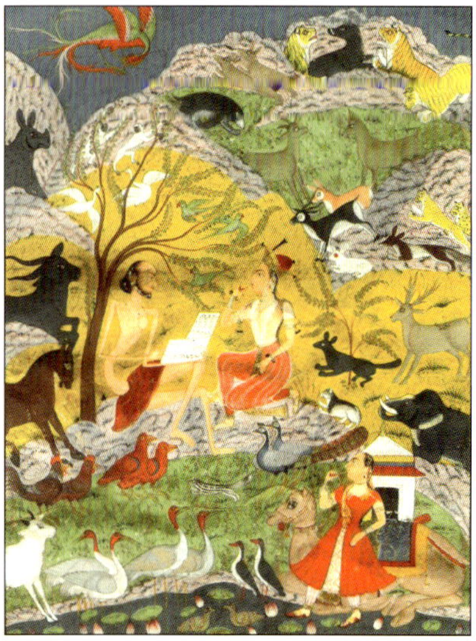

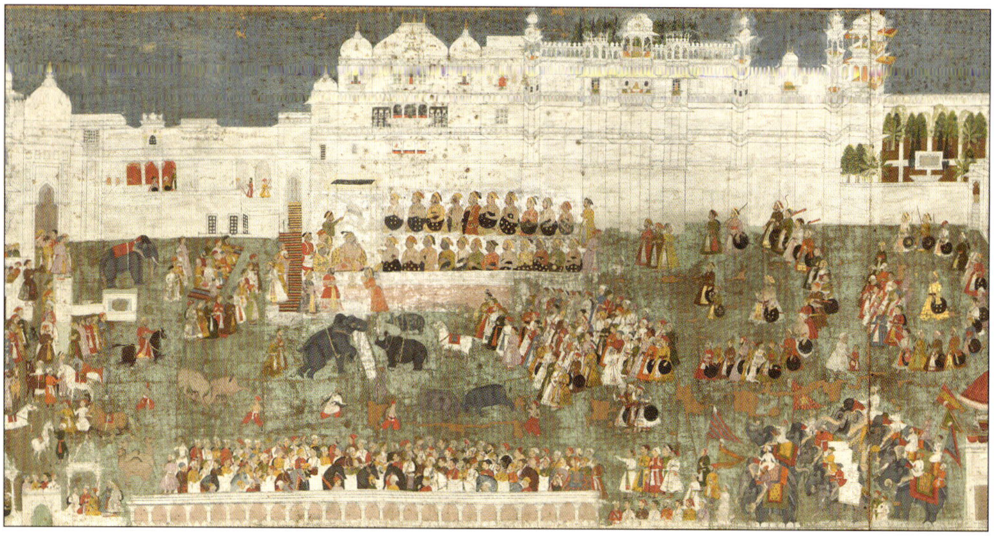

■ Layla meets Majnu in the wilderness, unidentified Rajasthani school, c. 1700. *Allahabad Museum.*

■ Detail of a painting of Maharana Amar Singh II and his court attending an animal fight in front of the palace in Udaipur, c. 1705. Painting on cloth. *Courtesy The City Palace Museum, Udaipur, © MMCF, Acc. No. 2012.20.0012_R.*

made in a Rajasthani atelier. Though the work has been attributed to Bikaner it seems stylistically closer to Kota or Bundi.[9]

The small picture of Maharana Amar Singh II of Mewar hunting a rhinoceros,[10] in contrast, looks stereotyped and lifeless. The ruler, seated on a huge elephant with an attendant wielding the flywhisk or chauri, seems to casually thrust his spear at the back of the animal, which is the least vulnerable area of its well-protected body. The third painting shows Maharana Jagat Singh of Mewar on foot, with an attendant carrying a spear, in a landscape before a huge rhinoceros standing on a wet grassy patch. It appears that the maharana is calmly chastising a tame rhinoceros.[11] The animal is not drawn from life, as is clearly apparent from its strange-looking snout, large round eyes and upright ears. The fourth work in the Allahabad Museum, showing the popular Layla meets Majnu in the wilderness theme, appears to be from the Bundi atelier. In this brightly coloured painting the rhinoceros is shown in a summary manner, entering the scene at top left to join the large number of birds and animals crowding the hilly landscape, including the mythical simurg in the upper corner.[12]

The drawing of a standing rhinoceros in the Chhatrapati Shivaji Maharaj Vastu Sangrahalaya, Mumbai[13] appears to be a 19th-century work from some Rajasthani atelier like Kota or Jaipur. The representation of the animal is very schematic. However, the finest depiction of the rhinoceros amongst this group of works is in the large court painting on cloth in the Maharana of Udaipur's collection. It shows a large gathering of men and officials before Maharana Amar Singh II.[14] They are about to watch a fight between a large elephant and a rhinoceros. The rhinoceros, as appears from the golden chain around its neck and a ribbon around its body, is trained for such a sport. Another smaller rhinoceros stands by its side.

As we have seen, it is known from travellers' records that the rhinoceros found place with other domesticated or tame animals in the menageries of the Mughal emperors. There is other evidence of such instances of the rhinoceros in captivity in writings and paintings. The well-known French traveller Jean-Baptiste Tavernier describes a tamed rhinoceros that he came across around 1665. He noticed a young boy feeding stalks of millet to a rhinoceros. When he took some stalks from the boy the rhinoceros came to him "opening his mouth four or five times" to eat the millet stalks.[15] Joachim Bautze discovered a late 18th-/early 19th-century drawing

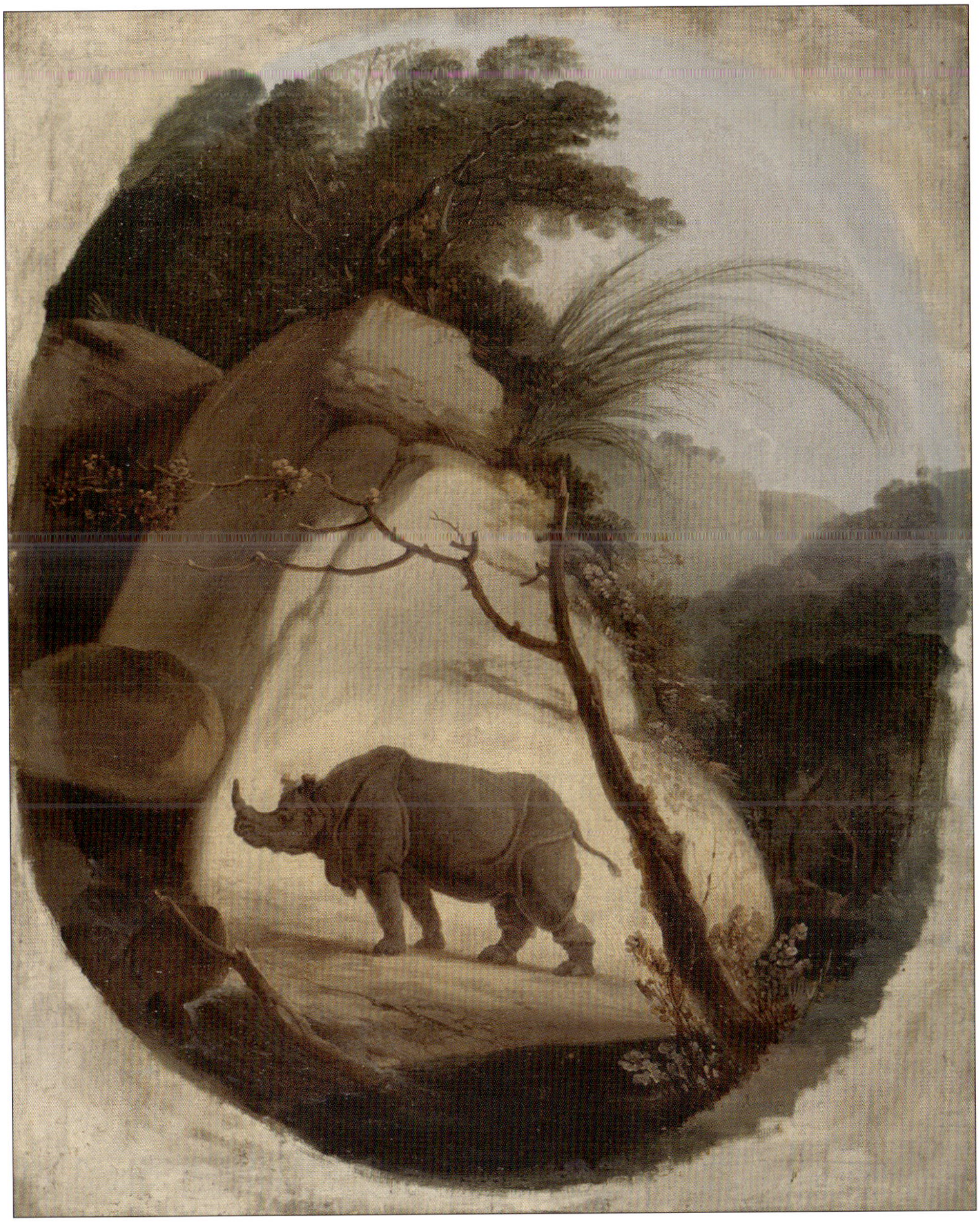

■ "The Indian Rhinoceros: A Forest Scene in the Northern Part of Hindoostan with a Rhinoceros", by Thomas Daniell, probably 1790. Oil on canvas.
Yale Center for British Art, B 2006-14-3.

showing two rhinoceroses chained to posts facing each other, in the Kumar Sangram Singh Collection, Jaipur. The drawing is not too inaccurate but the design of the armour-like plates on the animals' backs is rather schematic.[16] Further evidence of rhinoceroses being tamed comes from the accounts of the English Bishop Reginald Heber who made an exhaustive tour of northern and western India in 1824–25. He noticed several tamed rhinoceroses at Lucknow during his visit to that city. He also saw a rhinoceros in Baroda that was ridden by a mahout like an elephant.[17]

In the latter part of the 18th century the rhinoceros was found in large numbers in the eastern provinces, as revealed by an atlas prepared in 1769 by Colonel Jean-Baptiste-Jacob Gentil, the French representative to the court of Shuja ud-Daula at Faizabad, with the help of local painters like Nevasi Lal and Mohan Singh. The maps in the atlas show the topography and environment of the different provinces, and tiny figures of rhinoceroses are shown in the maps of Avadh, Bihar and Bengal.[18] Apart from the atlas, Gentil also commissioned paintings of court events, hunting

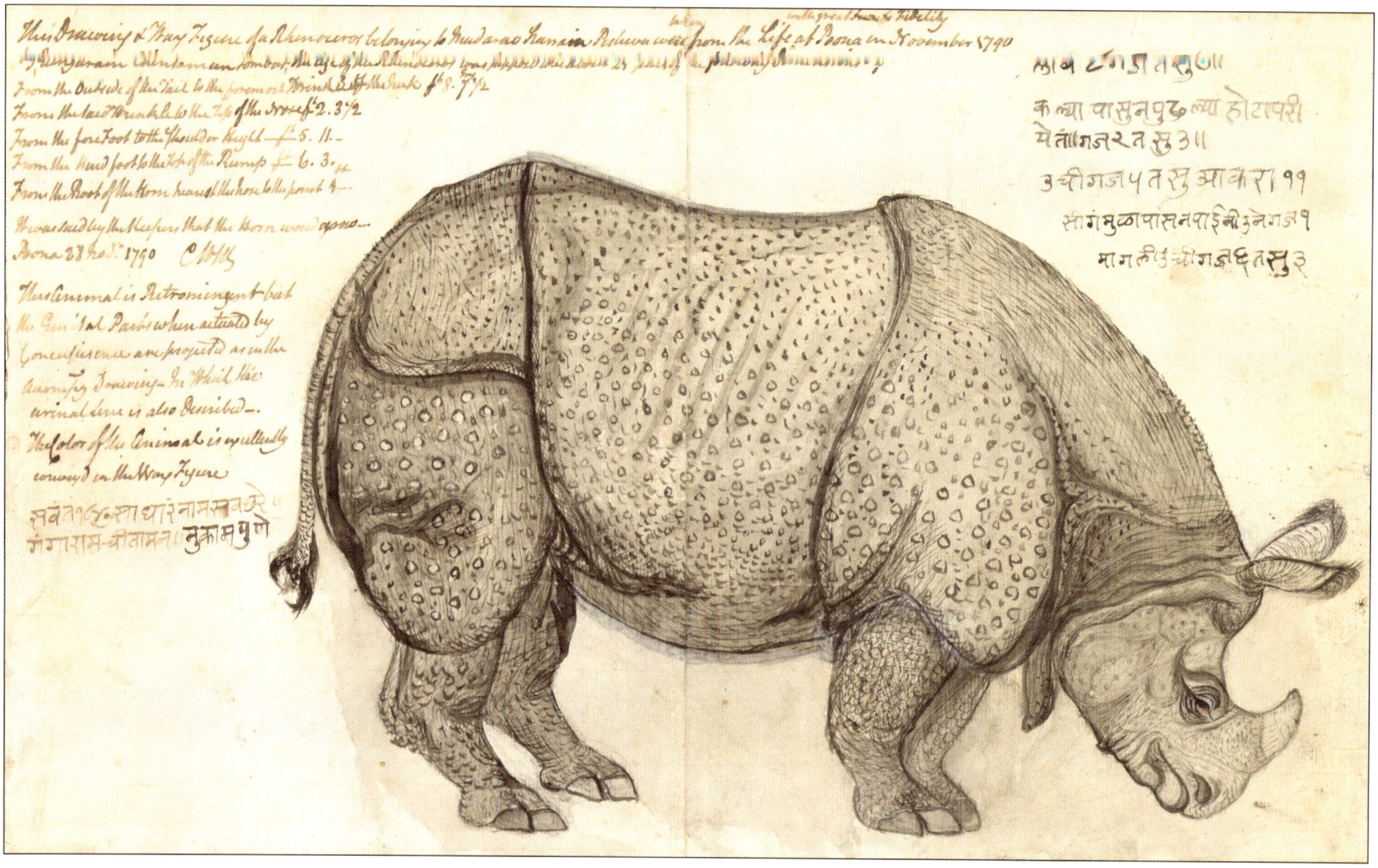

■ Rhinoceros from the Peshwa's menagerie, drawing by Gangaram Chintaman Tambat, 1790.
Yale Center for British Art, B 2006-14-33.

trips etc. which were mounted in albums. Two paintings from an album in the Victoria and Albert Museum collection provide elaborate details of two different hunting trips of the Nawab of Avadh, showing rhinoceroses amongst other big and small game. The first one shows a rhinoceros and wild buffalo hunt in 1768 in the woods of the Bahraich province of Avadh with a corps of 600 Frenchmen, among which Gentil was present. The nawab killed two rhinoceroses after they had wounded ten elephants. Apart from two dead and one wounded rhinoceroses it also shows a smaller one captured by the hunting party. The second painting shows another hunt undertaken by the nawab in the woods of Bahraich in 1769 that included Gentil and lasted for ten days. The painting depicts as many as 19 rhinoceroses, two lying dead, two fighting with elephants and as many as 15 large and small ones running away from the hunting party.[19]

Just a year later the well-known British landscape painter Thomas Daniell (1749–1840) noticed a male rhinoceros of excellent proportions in the foothills of the Himalaya and prepared a sketch from which he later, probably in 1790, made an oil painting (now in the Yale Center for British Art).[20] The horn of the rhinoceros drawn by Daniell however is longer than usual, like that of an African one.

Oddly enough the same collection also contains a superb drawing of a full-grown rhinoceros prepared by an Indian artist Gangaram Chintaman Tambat, who made many excellent drawings, paintings and clay models of people, birds, animals and landscapes for British patrons. This rhinoceros was in the menagerie of Peshwa Madhav Rao II at Pune, and the drawing contains the ruler's signature dated in

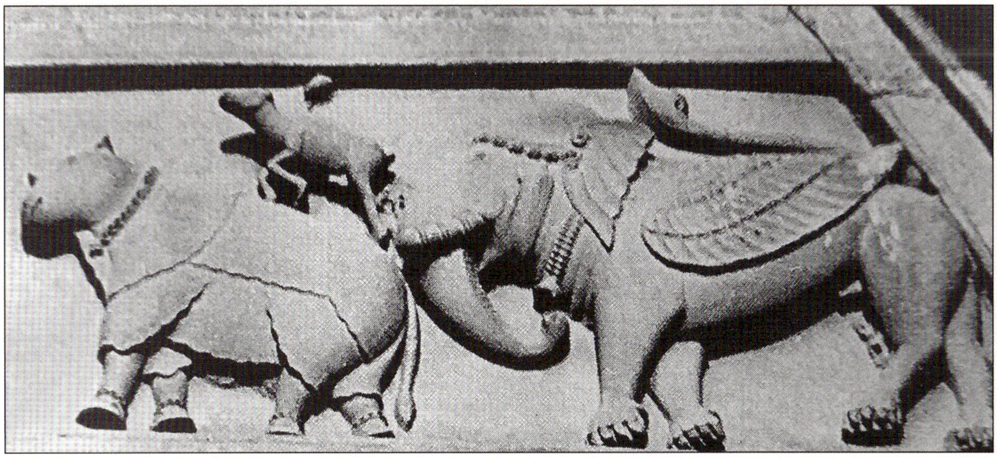

▪ Rhinoceros followed by a winged elephant composite, detail of a carved panel outside the western gateway of Sarai Nurmahal near Jalandhar, c. 1620.
Photograph: Dr Subhash Parihar.

▪ Europeans hunting rhinoceros, terracotta panel on the Gangeshvara Shiva Temple at Baranagar, Murshidabad.
Photograph: Asok Kumar Das.

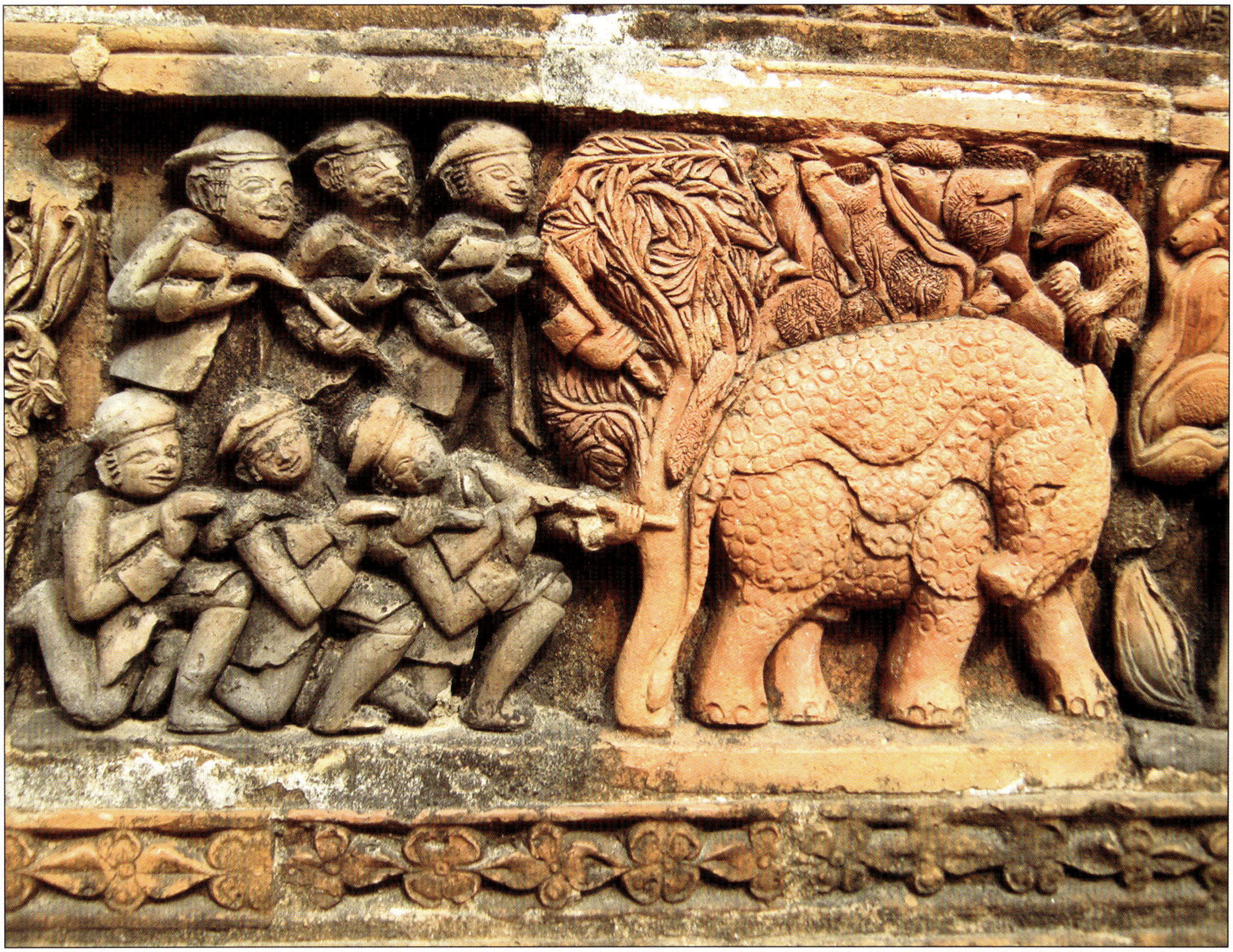

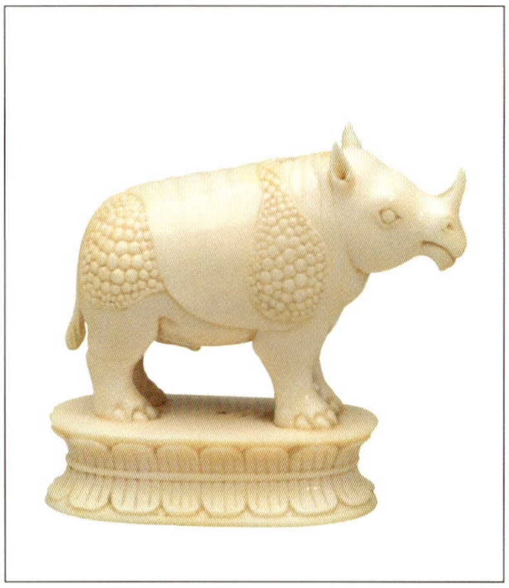

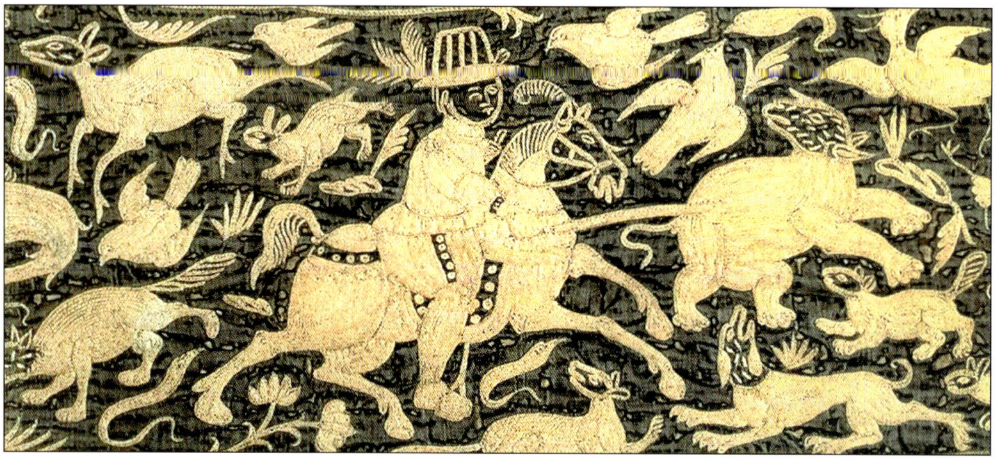

▪ Rhinoceros piece from a carved chess set, Murshidabad/Behrampur, c. 1820. Ivory and wood.
Courtesy of the Council of the National Army Museum, London, Acc. No. NAM. 1962-03-46-5.

▪ Detail from the border panel of an Indo-Portuguese quilt, Bengal, early 17th century.
Isabella Stewart Gardner Museum, Boston. Reproduced from Pedro Maura Carvalho, Luxury for Export: Artistic Exchanges between India and Portugal around 1600, *Boston: Isabella Stewart Gardner Museum, 2008.*

November 1790.[21] The drawing is of excellent quality, coming from an experienced painter and drawn from life.

Architectural Decoration and Artefacts

The image of the rhinoceros appears in other media as well. Subhash Parihar noticed several panels on the imposing western gateway of the sprawling garden and sarai named Nurmahal built by Jahangir, that included a mythical elephant-headed winged animal and rhinoceros together.[22] The winged animal is appropriately ornamented, while the rhinoceros is also shown with a metal chain around its neck and ornaments on all four legs. This is perhaps the only place where the rhinoceros is present in Mughal architectural decoration. Joachim Bautze noticed a beautiful sculpted marble panel in the Bada Mahal of Kota Palace with two very realistically rendered rhinoceroses facing each other along with other animals.[23] He also noticed rhinoceros representations on wooden doors at Kota.

The moulded or modelled terracotta panels in the brick temples in the countryside of Bengal, built in the 17th to 19th centuries, show rhinoceroses in hunting scenes. In a prominent panel in the lower portion of the Gangeshvara Shiva Temple at Baranagar on the western bank of the Bhagirathi near Murshidabad a full-size rhinoceros is shown under attack from a group of gun-toting foreign soldiers. Similar scenes with rhinoceros have been found in the Krishnarai Temple (1655), Bishnupur, the mid-19th-century Govinda Temple, Puthia, Rajshahi district, Bangladesh and the Kantaji Temple (1704–22), Kantanagar, Dinajpur district, Bangladesh.[24]

Among carved depictions of the rhinoceros, an interesting example is from a chess set made in Behrampur or Murshidabad, c. 1820, now in the National Army Museum, London. In this set, the pieces are of ivory, the two sides distinguished by their white or black bases, representing the British forces and an Indian ruler's army respectively. In keeping with the Indian tradition, the king and queen sit on elephants. The knights of the European version of chess are replaced by cavalrymen on the white side and camel-riders on the black; while the rhino substitutes for the white bishop and the buffalo for the black one.[25]

Textiles

Rhinoceroses are also found in textiles and carpets. In at least three early Indo-Portuguese embroidered textiles the rhinoceros is found amongst animals pursued by European hunters. These textiles made of cotton and tussar silk were produced

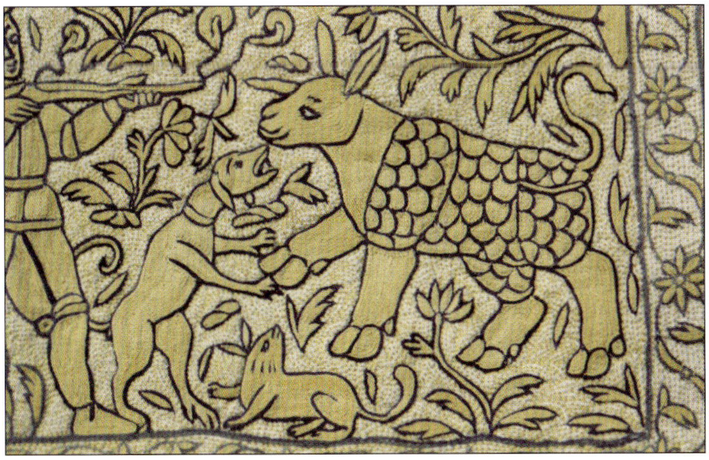

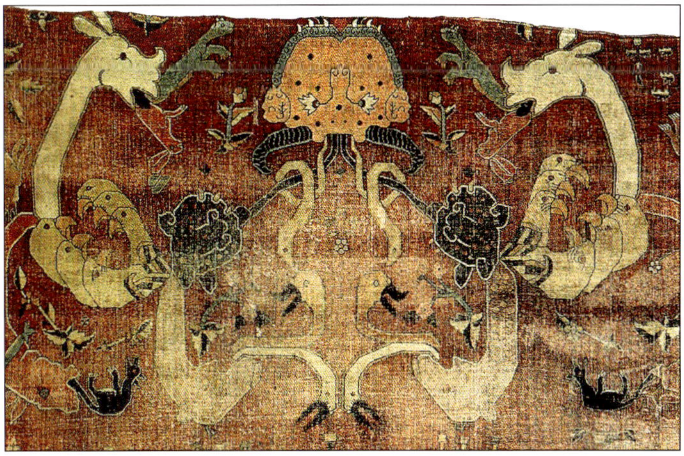

in the Satgaon area of Hooghly district in West Bengal exclusively for European clientele. The entire surface of the material was decorated with intricate silk-embroidered designs provided by the buyers. The design content often consists of Biblical themes, maritime scenes and scenes of hunting, expertly embroidered by the local craftspersons, mostly women.

A magnificent piece of an embroidered cotton and silk quilt in the collection of Isabella Stewart Gardner Museum, Boston, recently put on show in an exhibition at the museum, shows a European horseman pursuing a rhinoceros. The rhinoceros is clearly discernible amongst various birds and animals, real and mythical, that crowd the scene.[26] This was not entirely unexpected as rhinoceroses were found in large numbers in northern Bengal as well as in the Sundarbans deltaic area, and rhinoceroses were captured and transported to Portugal from the middle of the 16th century. A clearly drawn rhinoceros is found in the border of another large Indo-Portuguese quilt embroidered in tussar silk on bluish ground from the Museu Nacional de Arte Antiga, Lisbon, that was exhibited at Kensington Palace, London in November 1970.[27]

The rhinoceros also found a place in the magnificent pile carpets produced at the Mughal *karkhanas* at Agra and Delhi. The National Gallery of Art, Washington DC, has a large red-ground "animal carpet" with numerous birds and animals in the field and border design; a full-grown rhinoceros is shown in the upper part of the field.[28] In another carpet of rare and unusual design that survives only in small fragments in various museums and collections in the US, UK and Europe, rhinoceros heads have been used along with heads of various other birds and animals like elephant, tiger, cheetah, horse, deer, serpents, water birds, peacocks, and mythical birds and animals to form a complex "waq-waq tree"-like design in the field. The carpet (or a pair of carpets of similar design) is a unique creation of the Mughal studio, deriving its idea from the "waq-waq tree" or Tree of Alexander growing in the mythical island of Waq-Waq, that shows human heads or heads of birds, animals, real and mythical, dangling from its branches, as depicted in manuscript illustrations prepared in Persia, Mughal India and the Deccan. Elaborating on this, the carpet weavers (as well as the illumination specialists working for the embellishment of hashiyas or margins of folios of royal albums) created this harmonious as well as fearful design of birds, animals, aquatic creatures and serpents swallowing each other. The rhino heads depicted in the waq-waq design on the carpet fragment reproduced here are remarkably life-like and accurate.[29]

▪ Detail from an "Indo-Portuguese" quilt, Satgaon, Bengal, first half of 17th century.
Museu Nacional de Arte Antiga, Lisbon, No. 4574.

▪ Detail of a fragment of the "waq-waq" carpet, Mughal, late 16th century.
Burrell Collection, Glasgow, Inv. No. 1934/9/1.

■ Rhino-skin shield of Raja Chandrasinhji II of Wankaner state (1787–1809).
Courtesy M.K. Ranjitsinh.

■ Shield of Maharana Sangram Singh II from Udaipur, Rajasthan, c. 1730.
Courtesy National Museum, New Delhi.

■ Rhinoceros cup, Deccan, 17th century.
Kunsthistorisches Museum, Vienna. Reproduced from Exotica: The Portuguese Discoveries and the Renaissance Kunstkammer, *Lisbon: Calouste Gulbenkian Foundation, 2001, no. 59.*

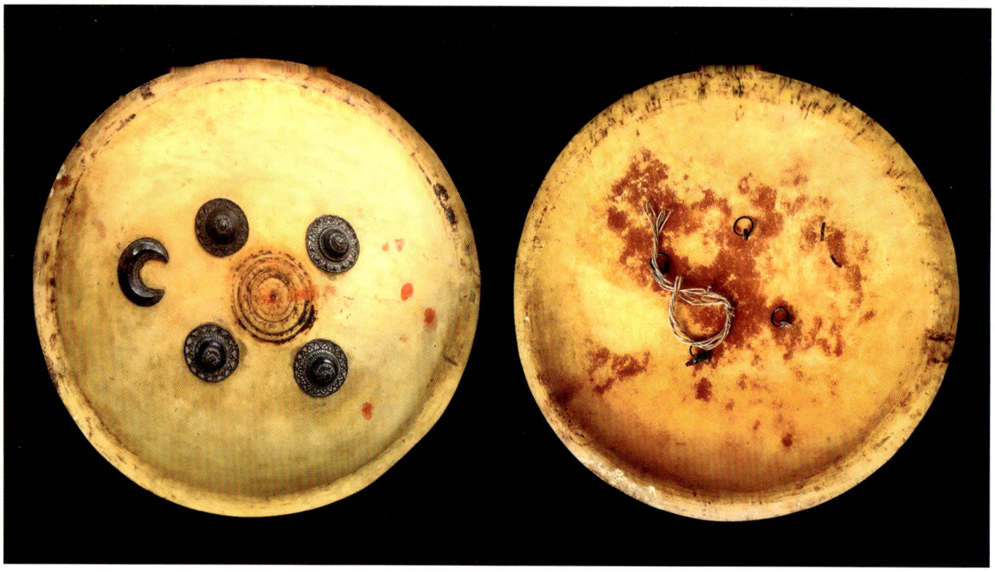

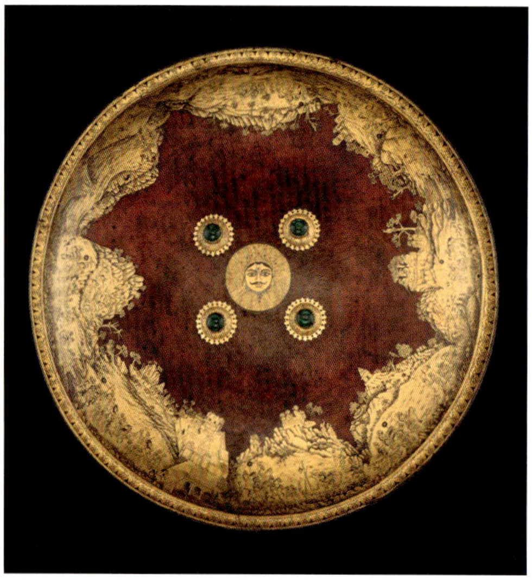

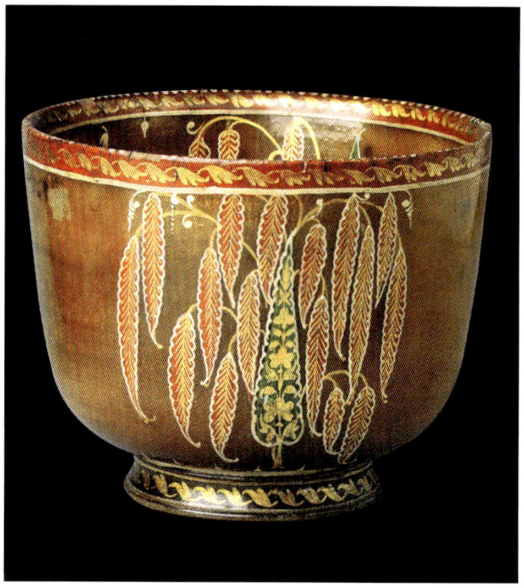

Objects Made from Rhino Horn and Hide

The use of the rhinoceros horn in India was not very widespread or popular. However the value of this product was never underestimated as it was always in high demand in China. The Chinese craftsmen created exquisite ritual objects, luxury items, goblets and other works of art from the Indian rhinoceros horn. References to horns of rhinoceros (*xijiao*) sent as gifts with the embassy from the Bengal Sultan Saif ud-Din to the Ming Emperor Yongle in June 1421 have been found in the Veritable Records of the Ming Dynasty, *Ming Taizong Shilu*.[30] Babur was aware of its use and a relatively unadorned but finely chiselled vessel in the Sloane Collection in the British Museum is generally regarded as the cup fashioned for Babur's use. However there is no firm evidence to support this association.

Abu'l Fazl in his encyclopaedic work on Akbar's reign, the *A'in-i-Akbari*, while providing accounts of the sarkar (fiscal unit) of Sambal (Sambhal) writes, "There is game in plenty in the Sarkar of Sambal, where the rhinoceros is found. It is an animal like a small elephant, without a trunk, and having a horn on its snout with which it attacks animals. From its skin shields are made and from its horn, finger-guards for bow-strings and the like."[31]

Shields and body armour made of rhinoceros hide are on display in the armoury or Silleh Khana of the Maharaja Sawai Man Singh II Museum, Jaipur. These were made in Kota and procured from there, as mentioned in old Silleh Khana records of Jaipur Durbar. Archer's rings for Mughal and Rajput princes and nobles, and earrings used by the Gorakhpanthis or Kanphata Jogis, followers of Gorakhnath, all fashioned from rhinoceros horn, have also been found in different museums.

The English traveller William Finch, who visited the Mughal court and parts of northern India in 1608–09, reports trade in objects like drinking cups, buckles etc. made from rhino horn at Ayodhya. Finch notes the high price rhinoceros horns fetched: "There are these hornes, all the Indians affirme, some rare of great price, no jewel comparable, some esteeming them the right unicorns horne."[32] The real and symbolic value of the rhinoceros horn is well demonstrated in an incident reported by Tavernier when he noticed how in a rhinoceros hunt around 1665 the animal's horn was cut off and the king presented it to the ambassador who was in the hunting party.[33]

There are numerous beautifully carved cups and decorative objects fashioned from rhinoceros horn in different museum collections throughout the world, most of them coming from Chinese workshops. Joachim Bautze has mentioned (on the strength of the evidence provided by the scholar Gouriswar Bhattacharya) the presence of a carved bowl made of "rhinoceros forehead" in the State Chandradhari Museum, Darbhanga, which shows, besides other figures, carvings of *Dasavatar* (the ten incarnations of Vishnu).[34] The sole rhinoceros-horn cup in the Indian Museum, Kolkata, probably of Indian origin, is well finished but unadorned. A superb footed cup made in the Deccan, now in the Kunsthistorische Museum, Vienna, carved from rhinoceros horn and with designs of trees and plants painted in colours and gold on its inner and outer surfaces, was shown in a 2001 exhibition in Lisbon of rare and exotic luxury objects from India, Africa and Sri Lanka.[35]

From this brief survey it is clear that the rhinoceros found place in the arts and crafts of the country from Rajasthan in the west to Bengal in the east. Though it never attained the widespread representation and popularity of the elephant or lion, it retained its place as a quiet, almost docile, and easily recognizable animal.

▪ The only natural predator the rhino has to contend with is the tiger which would more likely go for its young. The mother keeps its calf under constant vigil, Kaziranga.

Photograph: Giovanni Mereghetti/Marka/Dinodia.

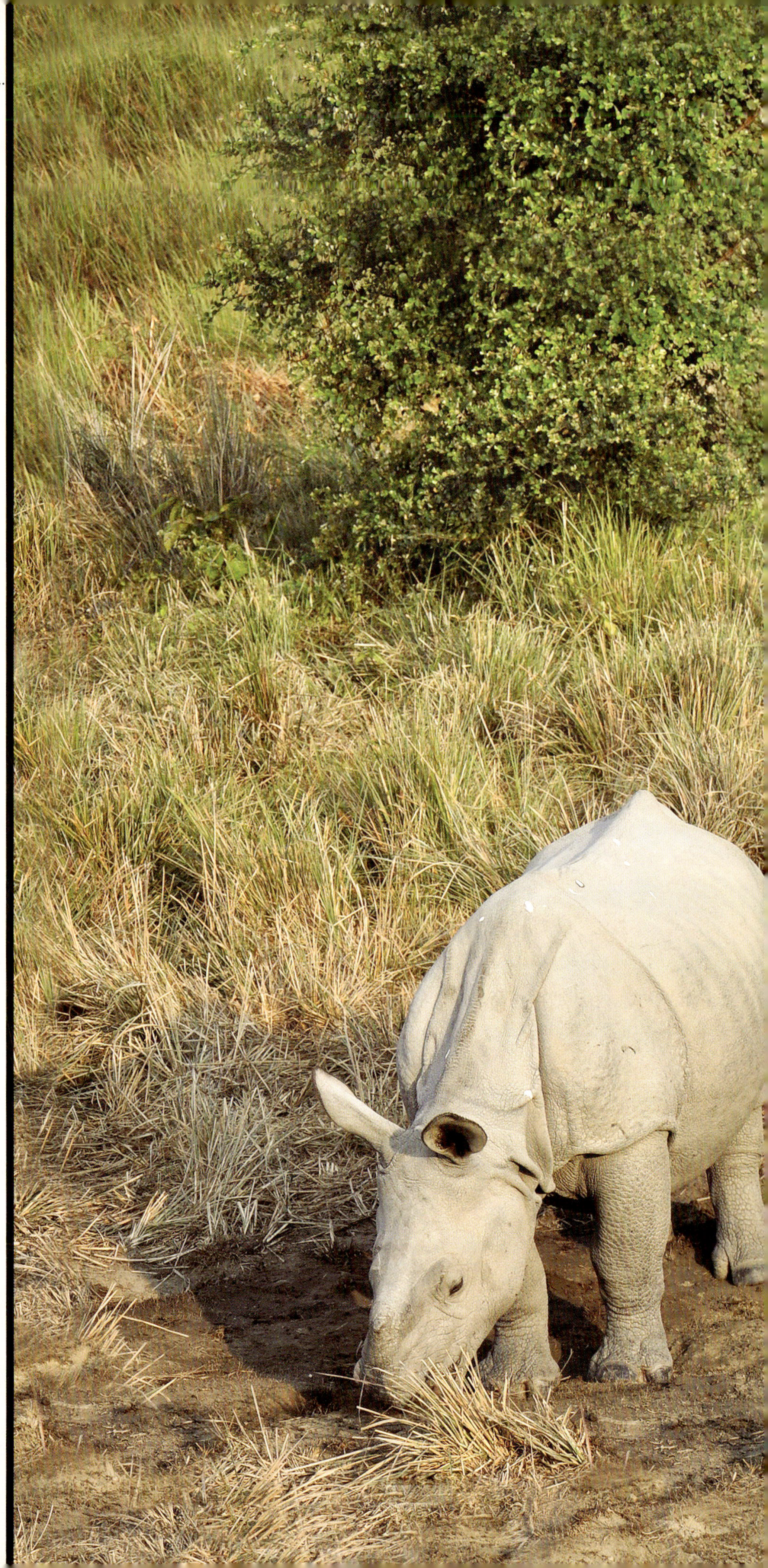

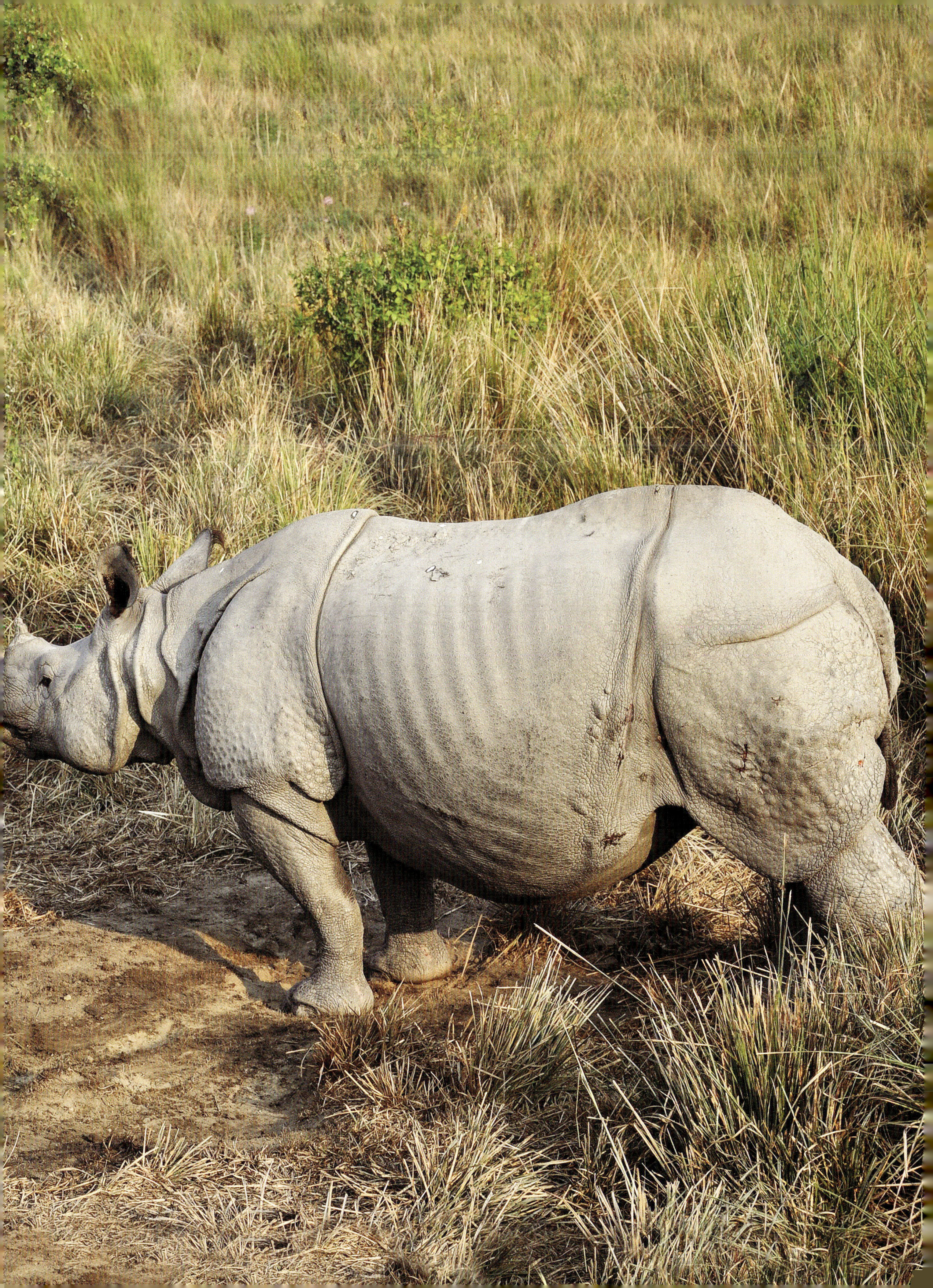

CHAPTER 6

Under the British and Beyond

Divyabhanusinh

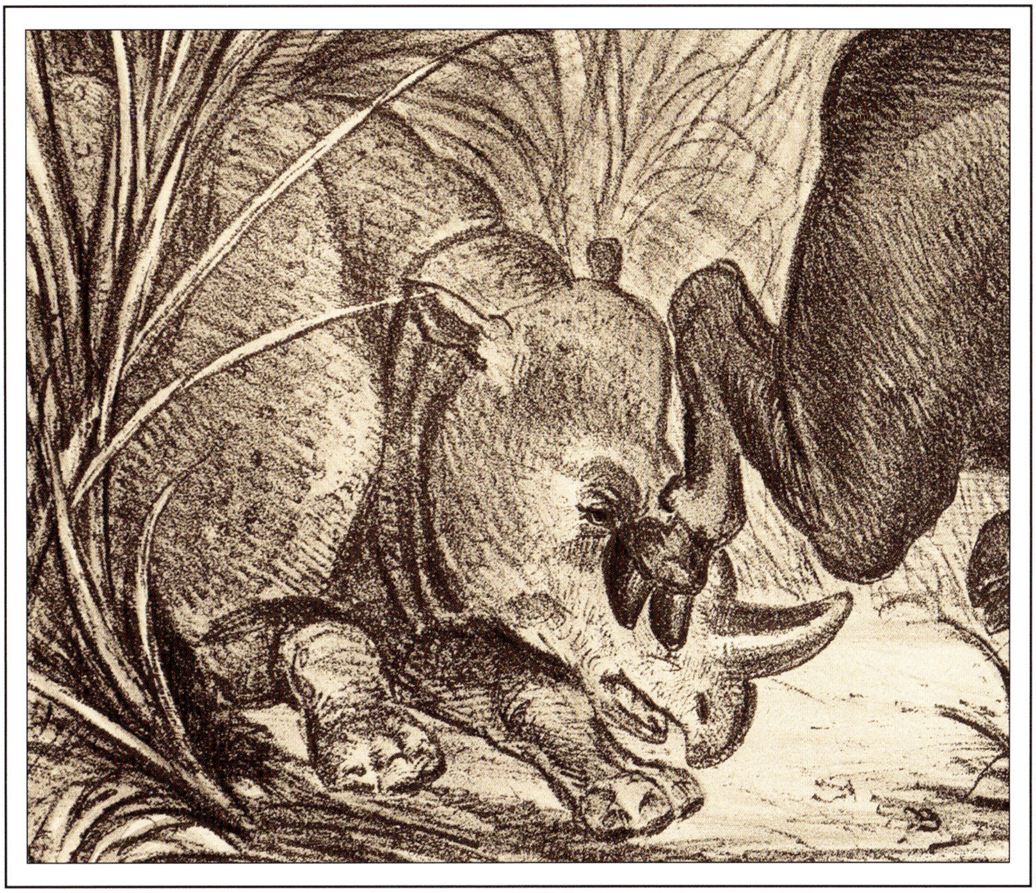

The last great Mughal emperor Aurangzeb spent the final years of his long life in the Deccan trying to subjugate the kingdoms of the south. With his death in 1707 signs of the fragmentation of the imperial order became clear. The empire became so weak that it could not stop the raid of Nadir Shah of Persia who sacked Delhi in 1739 and departed with much loot, including the priceless Koh-i-nur diamond and Shah Jahan's famous peacock throne. Still, Aurangzeb's successors largely managed to hang on to their ever-shrinking kingdom until Bahadur Shah Zafar, who was reduced to the title King of Delhi, was exiled in 1858.

Amidst this turmoil in 18th-century north India—the region that concerns us—the British East India Company went from being a trading company to become a political power. The Battle of Plassey in 1757 and the Battle of Buxar in 1764 made the Company the most powerful temporal power among the competing fledgling states that were flexing their muscles against the weakened authority at Delhi.

Field Sports under the British

The British who came to India as writers-clerks-boxwallahs-traders saw immense opportunities for relaxation from the drudgery of their daily routine in an inhospitable climate. Many who could never have dreamed of such opportunities "back home" took to a variety of "sports" with gusto and wrote about their exploits, shikar being the foremost among them. Books on the subject began to be written and continued to be published until India's independence in 1947 and after.

The nature of such publications is well analysed by Scott Bennett who writes: "The history of shikar has been left to be written by participants...we seek in vain for any analysis of what was done, and why it was done. In fact, the aim of such writers tended to be quite modest. To tell a good story, to inform the curious and to instruct the novice was all that was sought, and we are thus left with a remarkably uniform description of what was seen simply as an innocent amusement. A recent example of the genre is Major General James Elliot's *Field Sports in India 1800–1947* published in 1973.... The overwhelming impression gained from an account like Elliot's is that *shikar* was a harmless routine for the Britisher, which gave him (rarely her) a chance to break the bonds of civilization, while offering the chance of enjoying some of the better things of life."[1]

There were, as well, works such as William Rice's *Indian Game (From Quail to Tiger)* published in 1884 which informed the "griffin"—novice—of the type of sport available in a particular locality. Others like A.E. Stewart's *Tiger and Other Big Game* published in 1927 had chapter headings such as "Arrival at Your Block", "Tying out a Kill", "Inspection of Kills" etc. which speak abundantly of the nature of the contents.

It is in such texts that we look for references to the unicorn and its brush with *Homo sapiens* during this period. First off the block was Captain Thomas Williamson's *Oriental Field Sports* published in 1807, exactly half a century after the Battle of Plassey. He was in the Bengal Army, and lived in Cawnpore (Kanpur) and also in Calcutta which was the centre of the growing British Empire. Williamson was a colourful character. He took on the Governor-General Lord Wellesley, writing a libellous piece against him. He was cashiered and sent back to England in 1798 and there is believed to have penned his magnum opus to make some money. The book has a chapter on the rhino and a plate titled "Rhinoceros hunted by Elephants".[2] It is apparent from the text that Williamson had no personal knowledge of the animal

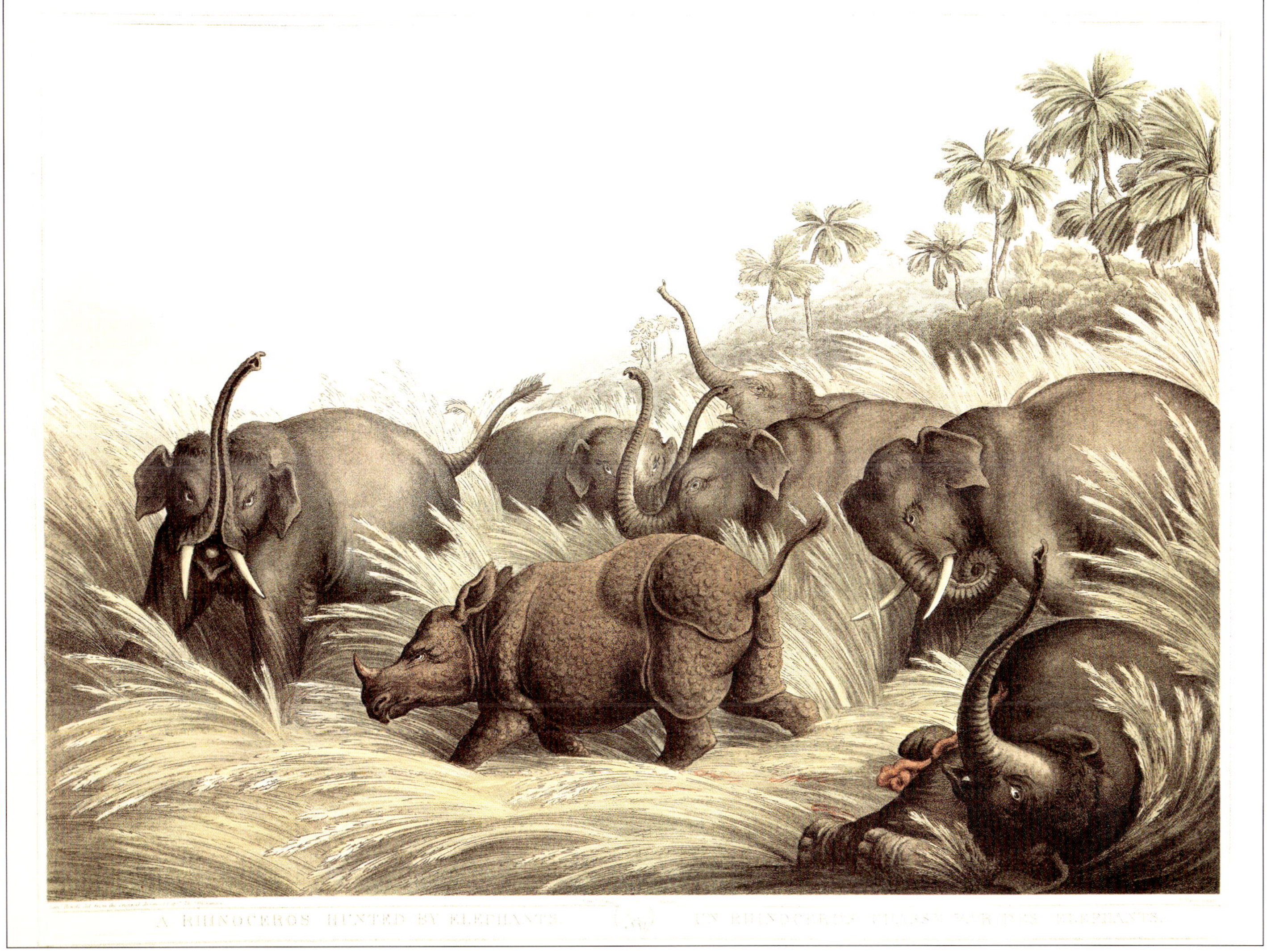

and merely recorded what he was informed by others, but his account is significant for being among the earliest of such records.

At the outset Williamson confesses that little is known about the rhinoceros's "habits, its powers, and many other very interesting points, [and] nothing authentic has been published". He goes on to describe the animal, but of greater interest is his record of fights between rhinoceros and elephant in the wild. He refers to a Major Lally who witnessed such a fight but did not see the end of it as the animals moved into thick jungle. Williamson also recounts an incident of a marauder rhinoceros encountered by two army officers at "Derriapore" near Monghyr in Bihar in 1788. The animal was reported to have attacked two horses of the officers' camp, goring them to death, thus proving "the cruel disposition of the animal".[3]

Robert Armitage Sterndale, writing nearly eight decades later, records that the rhinoceros is known to attack wild buffaloes and illustrates his point by a picture in one of his books titled "More than his Match".[4] It is obvious that there could be such conflicts among large herbivores who share the same habitat. However, different species usually manage to keep their distance from each other in the wild, and fights which may result in grievous wounds or fatalities are rare indeed.

Rhinoceroses were hunted by both Indians and Britishers in this period. The

■ "Rhinoceros hunted by Elephants", an illustration from the first book on shikar of the British period, Williamson, 1807. The depiction of the elephants is not entirely accurate.
From Williamson, 1807.

▪ "More than his Match": a rhinoceros drives off a wild buffalo from its territory.
From Sterndale, 1886.

▪ "Bull rhinoceros. A glorious prize." A rhinoceros shot by an indigo planter for whom it would have been a proud trophy.
From Inglis, 1888.

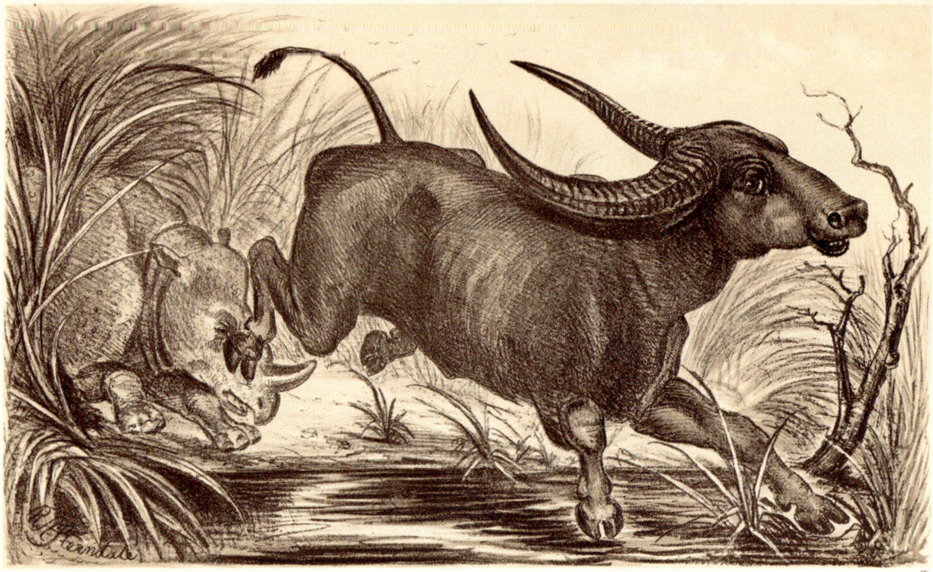

MORE THAN HIS MATCH

Published by Thacker, Spink & Co, Calcutta.

Page 298

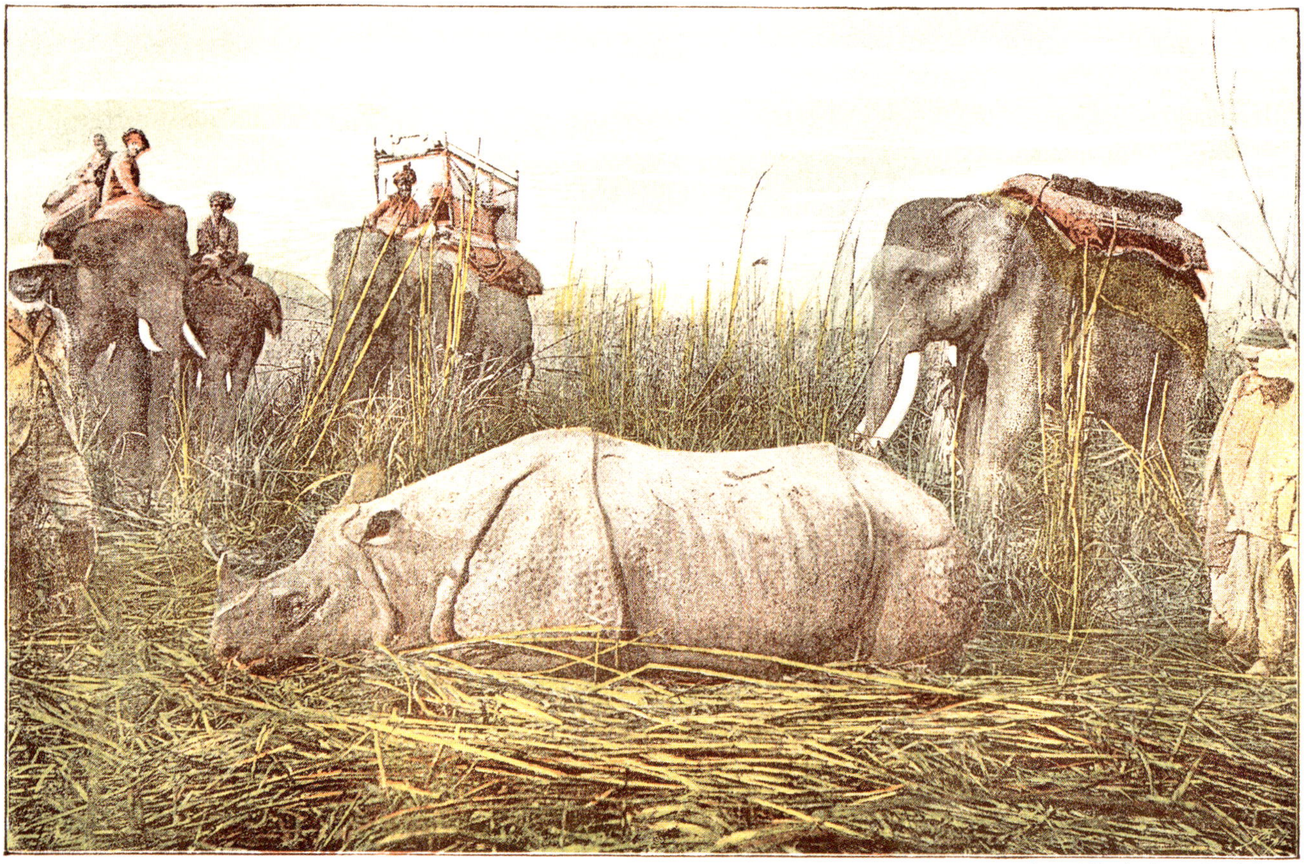

Vincent Brooks, Day & Son. Lith.

Bull rhinoceros. A glorious prize.

common people among the Indians used spears or muzzle-loading guns and the potentates among them used tame elephants to encircle an animal in tall grass before they finished it off with a gunshot. As for the British, they took to this "sport" of hunting the rhino enthusiastically, and usually hunted from elephantback.

Williamson's book was soon followed by others which also recorded the "sport". James Inglis was an indigo planter in northern Bihar on the Nepal border, who worked there just before and after the revolt of 1857–58. In his *Tentlife in Tigerland*, first published in 1888, he recounts that once when he was "on foot not thirty yards away but on the opposite side of a soft boggy ground...forth from the jungle in head long desperate flight came a magnificent full grown bull rhinoceros.... There was little time for reflection. It was evident that the angry brute had seen me, and with a hoarse, choking grunt of wrath and defiance he came plunging straight for me, rushing right into the morass." The animal got stuck in the bog, whereupon Inglis shot him broadside with two bullets, and gave a "yell of triumph, which I could not suppress" as "I saw the mighty brute sway to and fro, heaving his ponderous body as one may see a giant of the forest swayed by a rushing wind, and then with a hoarse groan he lurched forward struggled again through the tenacious clinging mud, and then crashed heavily over almost at my feet. What a glorious prize!"[5]

In an earlier work, *Sport and Work on the Nepaul Frontier* (1878), Inglis, writing under the nom de plume "Maori", records an incident when he came across a rhinoceros that had been shot by a zamindar. The locals were hacking the dead animal up for meat. He mentions that the skin from the underside made excellent riding crops. The horn was very precious too, since "when a child is to be born, the husband gets a horn to keep in the house which ensures safety to both mother and child".[6] That besides, the British also ate the meat. According to Alexander A. Kinlock, the animal's "meat is excellent eating and is not easily distinguishable from beef. The tongue which is very curiously formed is particularly good"![7]

Alexander A. Kinlock also described at some length the mode of hunting rhinoceroses. One method was to quietly track the animal on a single elephant "until he is at last found in its lair, or perhaps standing quite unconscious of danger". The other way was by beating the jungle with "a line of elephants, guns being stationed at the points where he [the rhinoceros] is most likely to break cover. It is essential to have beaters who have authority as both the mahout and elephants have the greatest dread of the huge beast who appears to be far more formidable than he really is. When disturbed he makes tremendous noise crashing through the reeds, and grunting and snorting with steam engine power, but unless driven to extremities by being hunted into a corner, I believe that it is but seldom that he will really charge home. I have not yet witnessed an instance of his doing so." Kinlock went on to state that if there were too many guns—hunters on elephantback—there would be wild firing. He noted that he used a .12 bore rifle.[8]

The name of Lord Curzon, Viceroy and Governor-General of India (1899–1905), is traditionally linked to rhinoceros protection. Once, when the Viceroy was hunting tigers in the Nepal Terai, he went on an impromptu rhinoceros hunt. He was there with a very small party and having shot a number of tigers had a few days to spare, so a request was made to the Nepalese ruler for shooting a rhinoceros. The Rana agreed readily, saying that "if the Viceroy could manage to take only a very small camp, and did not mind a rough journey, it might be arranged". Curzon went along

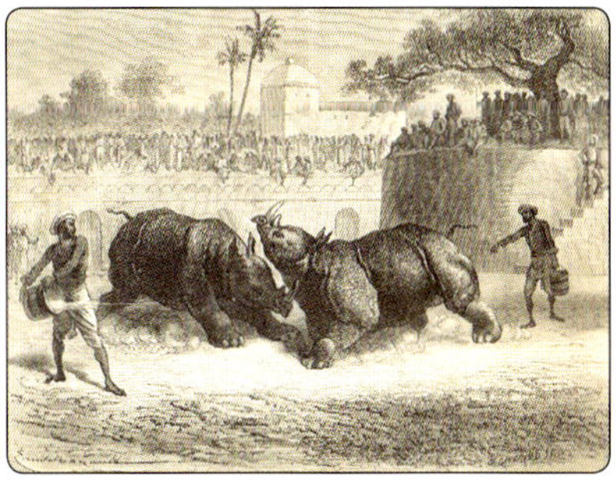

▪ A rhino fight at Baroda: rhinos of the royal menagerie were used for public entertainment in the 19th century. A fanciful representation as the rhinos here have two horns!

From Rousselet, 1882.

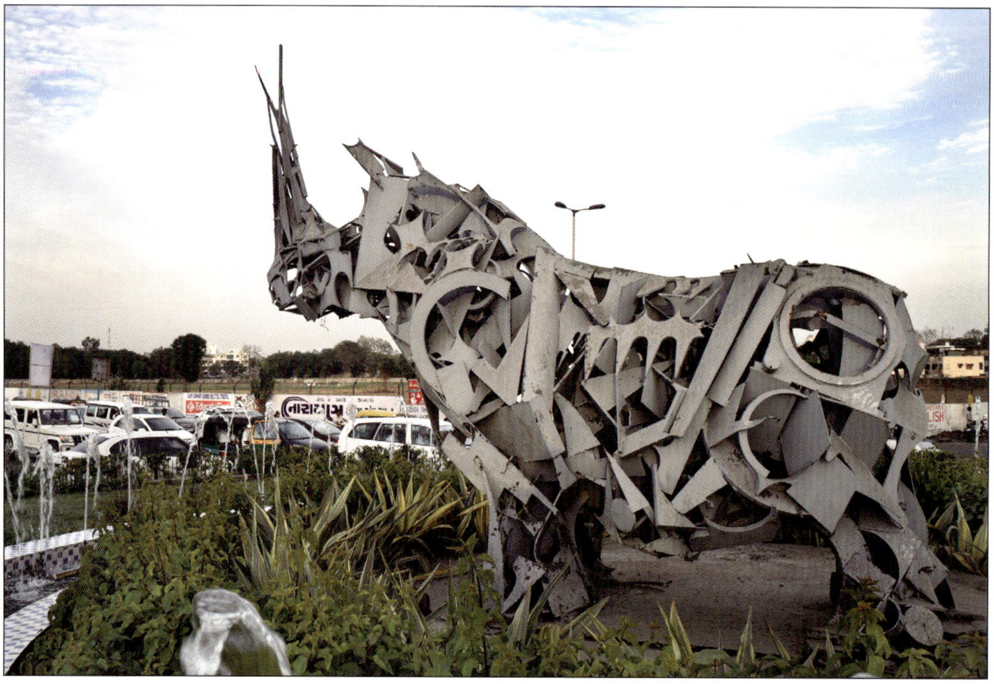

■ Genda Chowk (Rhino Square) in Vadodara is a reminder of rhinos in the royal menagerie and rhino fights of the 19th century.
Photograph: Manish Chauhan.

with only his aide-de-camp, Baker Carr. The area of the jungle to be disturbed by the beaters—people on foot who would make a noise to flush out the wild animals—was in a long riverbed which was 200 yards broad and 600 yards long. As the beat began, Curzon heard a sound like a railway train and saw a large rhinoceros charging towards him. He dropped it with a bullet of the .577 rifle that he had borrowed from his ADC.

All dismounted from their elephants, and the Nepalese beaters brought vessels to collect the blood of the animal oozing out of the wound as it was believed to have medicinal properties. While these proceedings were in progress the beaters suddenly shouted "Genera! Genera" (Rhinoceros! Rhinoceros!). The beaters scattered, the mahouts scampered up their elephants' trunks to their perches while the Viceroy and his ADC were hoisted into their howdah. Curzon managed to shoot this rhinoceros, which was the mate of the one just shot, before she could do any damage. Towards the end of the shoot a tiger also came out, which was bagged by Baker Carr. The horn of one of the rhinoceroses shot by Curzon was over 21½ inches long, which he believed for a long time was a record.[9] The record of course is that of a horn in the Natural History Museum, London with a length of 24 inches (see Chapter 1, page 23).[10] The hunter's prize usually ended up as trophy on the wall or an artefact or object of use such as a stool or a flower-pot.

There was yet another demand on the wild rhinoceros population. For centuries, they were captured and sent to menageries and zoos the world over. Kees Rookmaaker records that between 1409 and 1994, 2,439 rhinos of all species were in menageries; of these, 260 were greater one-horned rhinoceroses, and 137 were born in captivity.[11]

The Indian Princes and the Rhino

Kees Rookmaaker records that in 18th-century India rhinos were caught and semi-domesticated, sometimes even harnessed as draught animals to plough fields instead of buffaloes or bullocks.[12] However, the maharajas and nawabs had yet another use

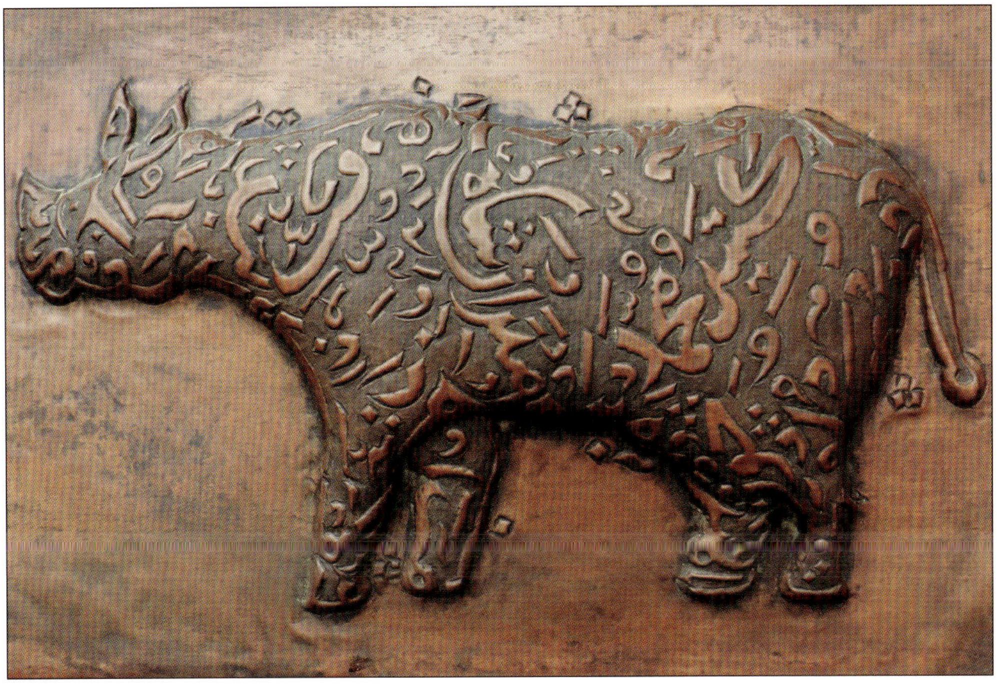

▪ Rhinos were part of Avadh's culture, as seen in this 19th-century image of an art form of this region. The Urdu letters read "Mohammed" and "Shah", possibly a reference to the contemporaneous Mughal ruler.
Courtesy M.K. Ranjitsinh.

for the animal. Besides being kept in their menageries, the precursors of modern-day zoos, fights were staged between captive rhinos, as indeed with other animals as well. In Baroda state (now in Gujarat), rhinoceros fights were a well-known event. To this day the city boasts of a "Genda Chowk" (Rhinoceros Square) though most people of the city have lost connect with the animal and the history of this gory spectacle.

By far the most famous of staged fights were in the kingdom of Avadh with its centres of Faizabad and, later, Lucknow (in today's Uttar Pradesh). Avadh was a *suba* (province) of the Mughal empire. The 18th and 19th centuries saw the central imperial authority disintegrating and the *subas* became independent, though the expanding reach of the East India Company was to soon curb their power as well. Avadh was ruled by Nawab Ghazi ud-Din Haider from 1814 till 1827, and it appears that fights between wild animals and beasts of prey were introduced during his reign, though such spectacles were long known at the imperial Mughal court. A special park called Hazuri Bagh was created along the river Gomti where such fights were staged, though smaller animals and birds were made to perform in the nawab's palace, Shah Manzil, itself.[13] Wild animals used in these contests included tigers, cheetahs, leopards, elephants and of course rhinoceroses. These were apart from fighting cocks, partridges, quail and other small animals and birds.

Abdul Halim Sharar, who chronicled the last decades of the kingdom before it was annexed by the East India Company, notes that in Lucknow, rhinoceroses were pitted against their own kind apart from having to tackle tigers, leopards and elephants. During Nasir ud-Din Haider's reign (1827–37) there were 15 to 20 fighting rhinoceroses in the menagerie. The keepers had to prod the animals to fight. When sufficiently angered one would go for the underbelly of the other and try to rip it open with its horn. They would bellow and lock horns, pushing against each other until the weaker one made a run for it. The victor would follow the beaten adversary and butt it with its horn. In larger arenas the vanquished could make good its escape, whereas if the arena was restricted the victor would kill its

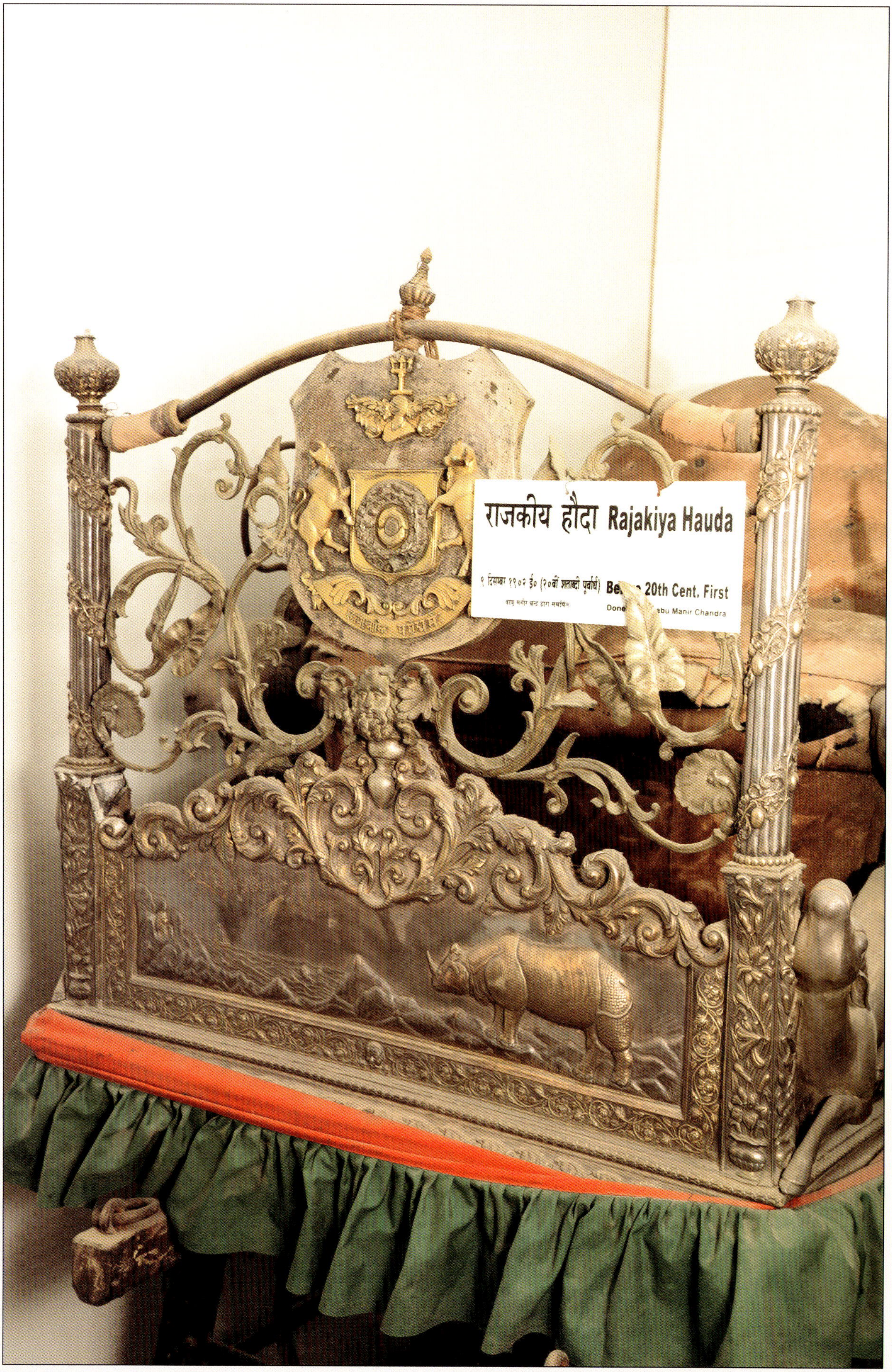
राजकीय हौदा Rajakiya Hauda

defeated opponent unless controlled by the keepers. Sharar goes on to record that the rhinoceroses had to keep their heads low to protect their bellies. If one made the mistake of raising its head, its opponent would take full advantage of this. It so happened on one occasion that a vanquished rhinoceros had started to run away, when the victor lifted its head; the other immediately turned about, lowered its head and ripped open its adversary's belly. In Lucknow some of these animals were also tamed enough to be harnessed to carts like elephants and they were even ridden.[14]

The Avadh tradition of rhinoceros menageries appears to have continued till 1847 when King Wajid Ali Shah was exiled to Calcutta after his kingdom was annexed by the Governor-General, Lord Dalhousie, into the possessions of the East India Company. It appears that a Javan rhinoceros went into exile with the king, and this animal landed up in the Zoological Garden at Alipore, Calcutta as noted in Chapter 1.

The figure of a rhinoceros in a landscape forms a unique decoration on the side of a howdah used by the Banaras maharajas on state occasions. Indian princes took to hunting the rhinoceros for sport. The most famous of these princes was Maharaja Sir Nripendra Narain Bhup Bahadur of Cooch-Behar, an avid shikari who spent his leisure hours with friends and relatives in pursuit of game. His principality in northern Bengal was an ideal habitat of the rhinoceros. On one shoot his party bagged five rhinoceroses before lunch. The maharaja published his memoirs of shikar expeditions between 1871 and 1907, during which period he shot 207 rhinoceroses, 365 tigers, 311 leopards, 48 gaur and 438 wild buffaloes.[15] While this was an extreme case, it does illustrate the enthusiastic pursuit of the sport by Indian princes, no less than that of the British.

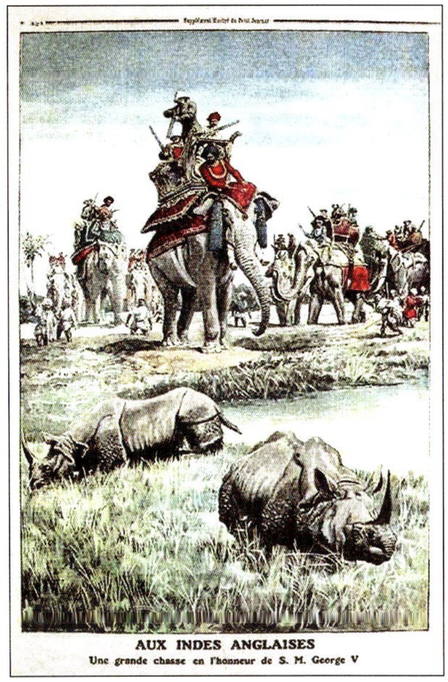

▪ A royal perch: the howdah of the maharajas of Banaras is unique in its display of a rhinoceros as part of its decoration. It has at various times carried several rulers, including Queen Elizabeth II, Reza Shah Pahlavi II of Iran and President of Yugoslavia Marshal Tito. *Courtesy Maharaja Anant Narain Singh of Banaras.*

▪ An artist's depiction of the royal shoot in Nepal in 1911. This illustration is from Le Petit Journal, Paris, December 31, 1911. *Courtesy Kees Rookmaaker.*

Declining Numbers

In addition to the hunting and capture of wild animals, human progress resulting in destruction of the habitat over the millennia took its toll. It is clear that the numbers of larger mammals decreased exponentially in the 18th and 19th centuries, and the rhino was no exception. By the beginning of the 20th century about 200 individuals survived, according to one estimate as noted in Chapter 1.[16] Numbers apart, chroniclers of shikar of the British period were certainly alive to the desperate situation that the animal faced and had started to express concern about it.

By the beginning of the 19th century Williamson had already noted that the rhinoceros was not found west of the Ganga. "Maori" recording his exploits later in the century noted that while rhinoceroses were not protected in the Terai area of India, they were protected by the Rana rulers of Nepal in their part of the Terai where outsiders did not have permission to shoot them. In fact when "Maori" accidently shot one there, his host was much exercised.[17] Captain J.H. Baldwin, a sportsman writing at about the same time, noted that the animal was found on the banks of the Sharda in Nepal, Pilibhit and Gorakhpur districts "but it is now extinct there or very nearly so".[18]

Alexander A. Kinlock noted that about the 1860s many rhinoceroses had been shot around Jalpaiguri, "but there partly owing to being constantly hunted, and partly owing to the clearance of large tracts for Tea cultivation, they are rapidly becoming scarcer.... In eastern portion of Bhutan Dooars and in Assam, wherever there are heavy reed jungles on the banks of rivers or on margin of swamps... occasionally several congregate in one covert."[19]

F.G. Aflalo's compendium of sport available to the British, written in 1904,

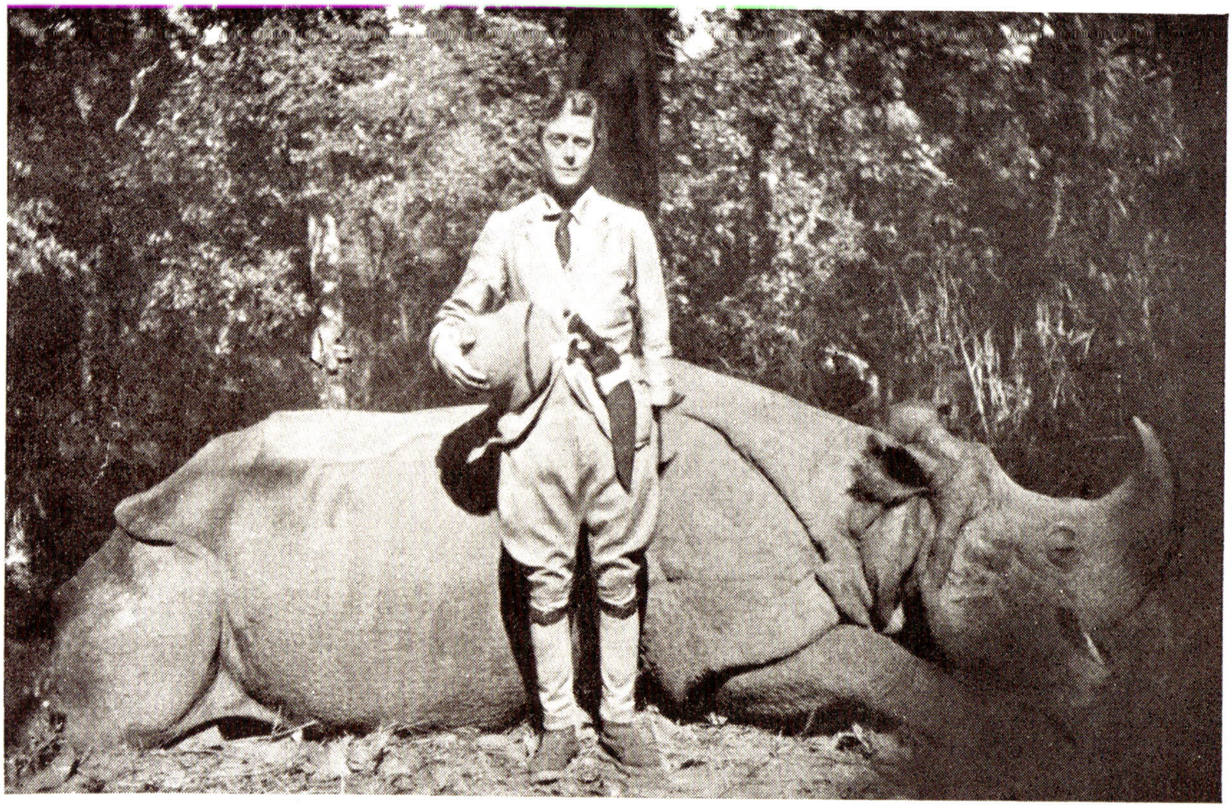

Photograph through the kindness of Colonel W. F. T. O'Connor, C.I.E., &c.

H.R.H. THE PRINCE OF WALES WITH HIS RHINOCEROS.
H.R.H. is wearing his Nepal kukri.

▪ Prince of Wales (later King Edward VIII) with his rhinoceros trophy in the Nepal Terai, winter of 1921–22.
From Ellison, 1925.

noted that 40 years earlier rhinoceroses had been extremely numerous and several could be hunted down in a single day. Aflalo went on to record that "Owing to indiscriminate slaughter of both sexes and all sizes, their numbers have been terribly reduced but there are enough left to enable a well equipped sportsman to be pretty sure of obtaining one or two specimen...." He also stated that shooting them in Nepal could only be done with a special permit.[20]

Writing in 1910, Sainthill Eardley-Wilmot noted that while rhinoceroses and bison (gaur) were found earlier in the Bengal Terai, they had now become very scarce and shooting them was prohibited. He wrote that the "Maharaja of Cooch Behar was a keen sportsman and the head of game had become insufficient to afford diversion both to himself and his guests...."[21] However, he stopped short of saying that the maharaja himself was responsible for this depredation!

E.P. Stebbing recorded in 1920 that the building of railways and the increasing number of sportsmen were the cause of the rhinoceroses decline. C.H. Stockley too noted the reduction in rhinoceros numbers in 1928 and went on to state that protection of the animal in Nepal had kept its numbers from being depleted in that kingdom. R.D.T. Alexander and A. Martin-Leake in *Some Signposts to Shikar* noted in 1932 that it was very difficult to obtain a permit to shoot a rhinoceros in India and that was "as it should be for any rare or near extinct animal".[22]

From these and other records of "naturalist sportsmen" it becomes evident that early on during the British period the rhinoceros was in near irreversible retreat. In India the colonial government had awakened to the problem by the early 20th century, as will be noticed from records after 1905 that rhinoceros hunting required a permit from the ruling authority in British India and in most of the princely states in India, whereas the Rana rulers of Nepal already had afforded protection to preserve the object of their most prominent sport. The reasons for the protection of the animal may have been dubious; the result however, was welcome.

Protecting the Rhino

It is evident that the battle for the protection of the rhinoceros was to be fought in the easternmost regions of India, in the Assam region of the Bengal presidency. The issue of depleting numbers of the animal was recognized by the administration as early as 1902 when J.C. Arbuthnott, Officiating Commissioner of the Assam Valley districts, wrote to B. Fuller, Chief Commissioner of Assam, on November 4 stating that the animal had been wiped out except in the remote localities of Bhutan Hills in Kamrup and Goalpara and between the Brahmaputra, Mikir Hills in Nowgong and Golaghat "where a few individuals still exist". Fuller replied on December 18 that while he regretted the plight of the animal it would be impossible to penalize unlicensed shooters without special legislation though he was open to create an asylum for them in the form of a reserved forest.[23]

The possible areas for such a reserve were surveyed by an army officer, a Major P.R.T. Gurdon, and on September 20, 1904 E.C. Carr, the Conservator of Forests, recommended that Kaziranga be declared a Reserved Forest. This was promptly approved by Fuller and on June 1, 1905 a Gazette notification proposing to declare it as such was made. On January 3, 1908 Kaziranga was declared a Reserved Forest and then upgraded to a Game Sanctuary in 1916. It appears that the Kaziranga area had become a haven for hunters and poachers, and hunting once again became rampant in the 1930s. In 1938 A.J.W. Milroy, Chief Conservator of Forests, opened the area to the public.[24] Until the dawn of India's independence, this position continued. The rhinoceros thus received protection in Assam.

Tradition has it that Lord Curzon was responsible for saving the rhinoceros. However, the facts brought out by Ranjit Barthakur and Bittu Sahgal conclusively show that these first steps were taken by a few determined, enlightened officers of the British Indian government. Barthakur and Sahgal record that the late S. Deb Roy, a remarkable Forest Officer of Assam and a conservationist, stated that Lady Mary Curzon, the Vicereine, had visited the Kaziranga region at the turn of the 20th century with the specific purpose of seeing this magnificent animal. However, she was not lucky, for all she saw was its spoor, and this led to her advocating for the protection of the rhino. There is yet another story doing the rounds: Lady Curzon went to Kaziranga in January 1905. She was accompanied by "Nigona Shikari", Bopiram Hazarika of Bosagaon, who convinced her that the animal she had seen in the distance was a rhino and not a buffalo. He showed her the distinctive tracks and urged her to stop the "Sahibs" from shooting them. There is, however, no written record of this visit[25] and the story is probably apocryphal.

There is no record of Lord Curzon visiting Kaziranga either, but, he was very much alive to the problem of conservation of wild animals. In 1901, the Burma Game Preservation Association sent a memorial to the Viceroy seeking the imperial government's approval to shoot thamin, the brow-antlered deer of Burma. In his celebrated reply of 1902, the Viceroy presents a tour de force of an argument for conservation that was rare in the empire at the time. His own words ring true even today: "There are some persons who doubt or dispute the progressive diminution of wildlife in India. I think that they are wrong. The facts seem to me to point entirely in the opposite direction. Up till the time of the Mutiny lions were shot in Central India. They are now confined to an ever-narrowing patch of forest in Kathiawar. I was on the verge of contributing to their still further reduction a year ago myself: but fortunately I found out my mistake in time, and was able to adopt a restraint which I hope that others will follow. Except in Native States, the Terai, and forest reserves, tigers are undoubtedly diminishing. This is perhaps not an unmixed evil. The rhinoceros is all but exterminated now in Assam. Bison [gaur] are not numerous or so easy to obtain as they once were. Elephants have already had to be protected in many parts...."[26]

Lord Curzon mentions specifically that he did not shoot a lion because their numbers had depleted drastically. He mentions that the rhinoceros had disappeared everywhere save in Assam. Yet he does not mention any personal intervention he made to save it, or that he had visited Kaziranga. He of course, was in office for three years more after the letter was written and he could have intervened during that time. Be that as it may, 2005 saw the century celebration of Kaziranga National Park and Sir Nicholas Moseley, the scion of Curzon's illustrious family, was invited to receive the *Sentinel of Kaziranga* tribute.[27]

What is absolutely clear though, from various reports of "naturalist sportsmen", is that permission to hunt rhinoceroses was difficult to come by in the 20th century, as noted earlier. The province of Bengal went ahead and formalized protection by passing "The Bengal Rhinoceroses Protection Act" in 1932 when the shooting of an animal without permission or sale of its body parts invited imprisonment for one month and/or a fine of Rs 1,000. This was followed by the Assam Rhinoceros Preservation Act of 1954 after independence which was on the lines of the earlier Bengal Act of 1932.[28] Thus the animal came to be legally protected throughout its remaining range.

The situation for the rhinoceros remained unchanged for the last two decades of British rule in India when the government was preoccupied with India's freedom struggle and then the Second World War. The rhinoceros was already in precipitous decline when the British arrived in India, a situation that was exacerbated as a result of the depredations by the British and Indians alike. When the British left, the animal was making its last stand, thanks to the steps taken by certain British administrators.

After Independence

The question, therefore, arises: what was the rhinoceros inheritance that the Republic of India received? The Javan and Sumatran rhinoceroses had long gone from the Indian subcontinent. What we know about the numbers of the greater one-horned rhinoceros is essentially a result of intelligent estimations by wildlifers, naturalists and forest officials, as no systematic population evaluation was done. That the situation was critical is obvious from the fact that as early as 1954 M.R. Medhi, Chief Minister of Assam, wrote to Prime Minister Jawaharlal Nehru that the rhinoceros was on the verge of extinction in his state.[29]

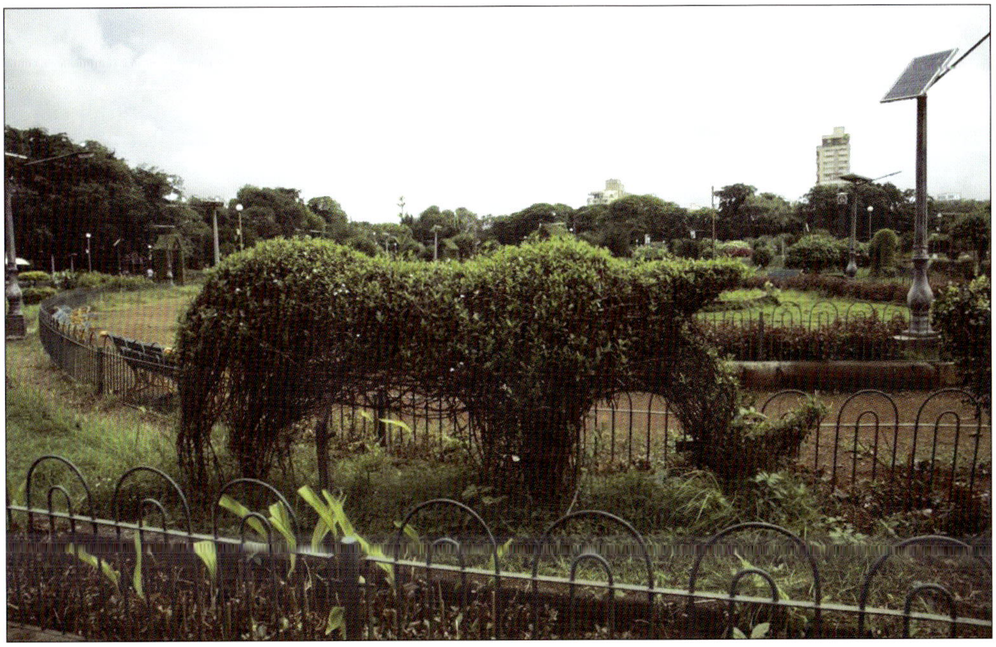

■ A floral tribute: rhino sculpture (topiary) at Hanging Gardens, Malabar Hill, Mumbai.
Photograph: Nirlek Dhulla.

P.D. Stracey, a well-known Forest Officer and Director of the Forest Research Institute and Colleges, Dehradun, gives us an early estimate of rhinoceros numbers. According to him in 1960 there were about 150 of them in Kaziranga Sanctuary, an area of 430 sq. km, out of a total of about 300 in Assam, with another 50 in Bengal and about 300 in Nepal. According to him the total population was likely to be over 600. But, he adds: "No accurate census of this animal [was] possible however and all estimates of its numbers are to be treated with caution."[30]

E.P. Gee, the well-known tea planter and wildlife conservationist, estimated rhinoceros numbers as follows: Nepal 185, Bengal 65, and Assam 375, a total of 625 in 1964.[31] Five years later, Balakrishna Seshadri, an engineer-turned-wildlife enthusiast, gives a figure of 745.[32]

It is evident from these records that only rough estimates were made, as none of the three gives us the basis of their figures. However, if one were to consider Kees Rookmaaker's estimation that there were about 200 rhinos in India around 1900 then one can conclude that years of conservation efforts by the British and Indian administrations were bearing fruit and the animal was making a bid for recovery by the end of the 1960s. This is on the assumption that the three estimates provide an indication of the general trend of the population dynamic, though poaching was a constant threat as it continues to be today.

In 1969 the Indian Board of Wildlife set up an "Expert Committee Re: National Parks, Wildlife Sanctuaries and Policy for Wildlife Conservation in India" under the presidentship of K.S. Dharmakumarsinhji of Bhavnagar. The Committee visited Kaziranga and Manas. They found both of these areas fit for being declared as National Parks which would give stronger legal basis for protection to their flora and fauna, the rhinoceros being among them. They also suggested strong conservation measures for Jaldapara in West Bengal.[33] Both Kaziranga and Manas became National Parks, with areas of 884.43 sq. km and 500 sq. km respectively, providing more secure refuge for the rhinoceros and other wildlife of the region. The animal itself went on to find a place in Schedule I of the Wildlife Protection Act of 1972, thus acquiring the highest status for protection.

▪ Rhino medallion: Nepal's iconic megafauna features on this issue to celebrate the first decade of King Mahindra Trust for Nature Conservation in 1994.
Courtesy Divyabhanusinh.

▪ On India's currency: the rhino appears on a 25-paise coin, now rarely seen, and with other mega fauna on the 10-rupee note (opposite).

▪ In India Postage: the rhino adorned an early postage stamp (opposite), and in 2016 the postal department issued a first-day cover at the behest of World Wide Fund for Nature–India (below). The postage stamp on the first-day cover depicts a black rhino from Africa.
Courtesy Divyabhanusinh.

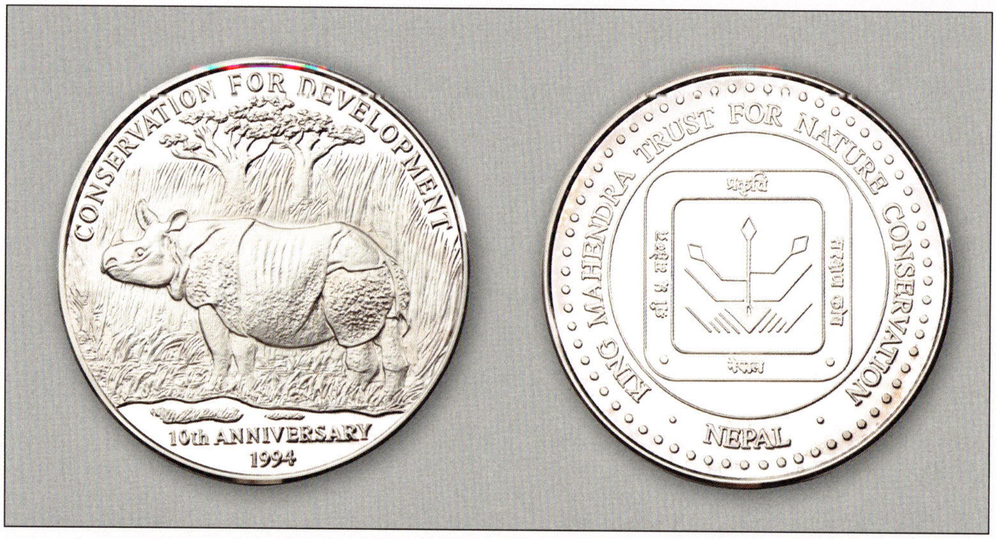

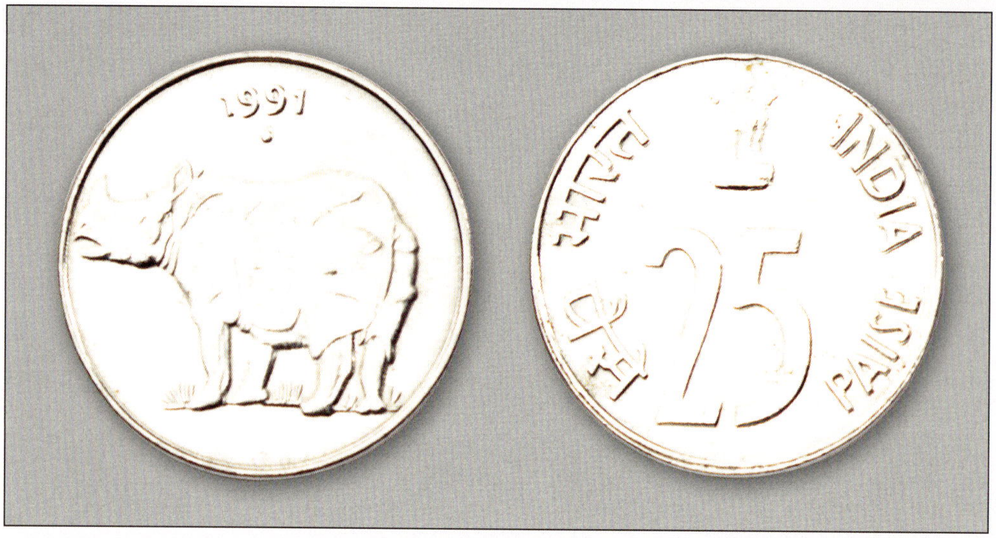

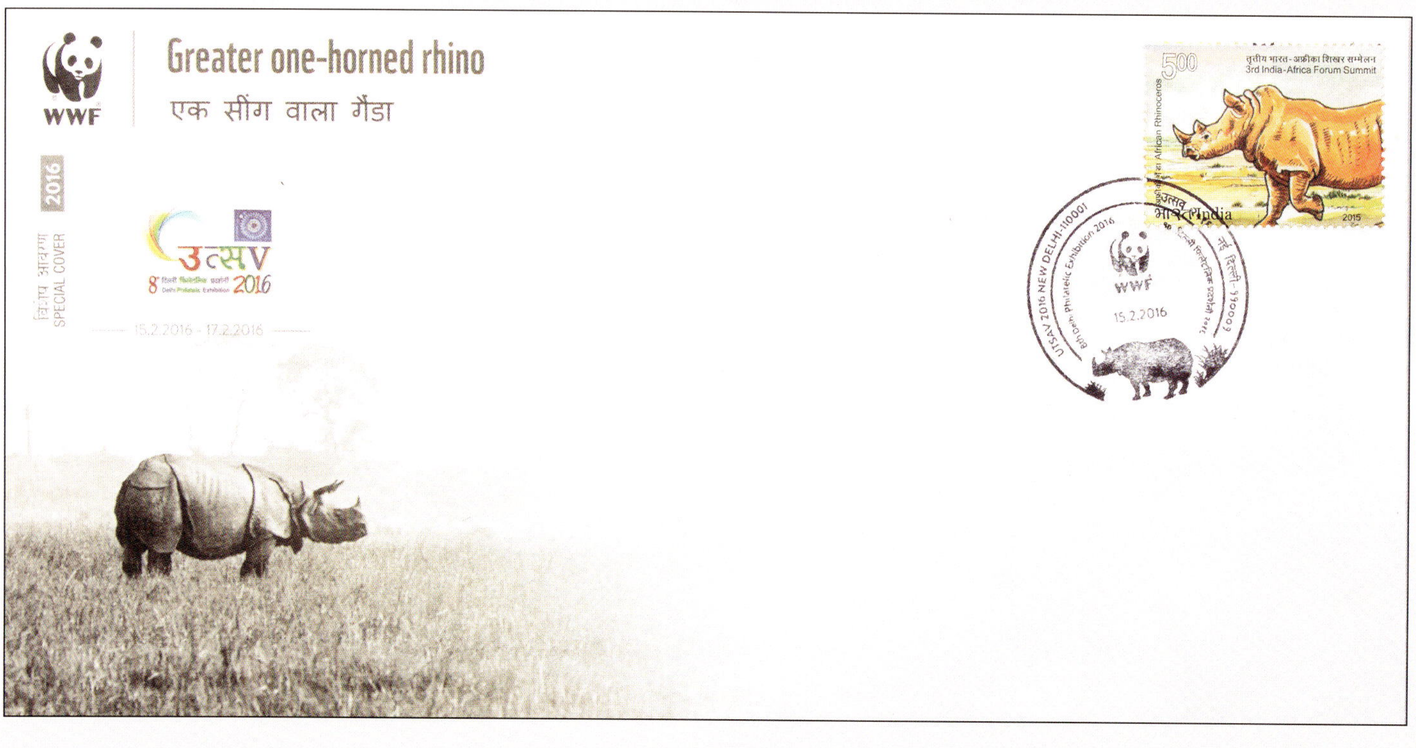

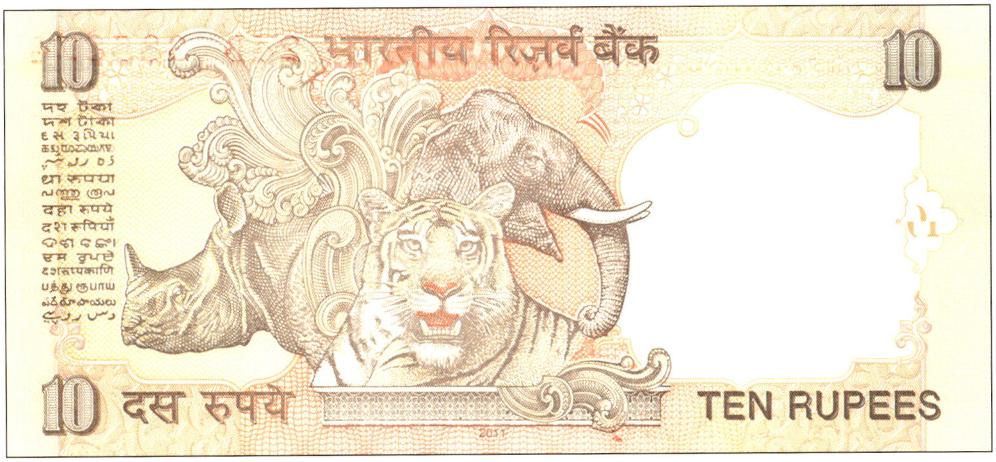

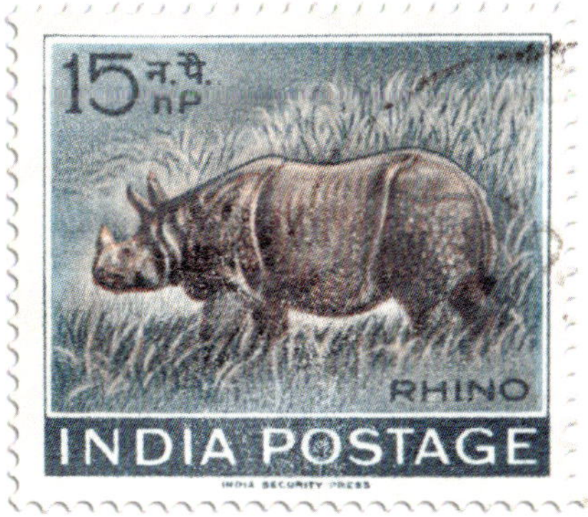

In addition to the government's efforts, there is another reason for the gradual success of the protection of the rhinoceros over the last several decades. The Assamese people take a special pride in their "gainda". It has a unique place in local tradition. According to one legend the Princess of Sonitpur, daughter of King Ban, had a dream of a handsome prince with whom she fell in love. Her friend Princess Chitralekha drew likenesses of all the eligible princes of India until finally she drew a picture of Aniruddha, the grandson of Lord Krishna, whom the princess recognized as her dream prince. Aniruddha came to Sonitpur to fetch the princess and apparently he was confined there by the king. Then Lord Krishna arrived, riding on a rhinoceros all the way from Dwarka, to rescue Aniruddha. He left his mount in Kaziranga to graze and crossed the Brahmaputra to wage war. Upon the successful completion of his mission he played his flute to call the rhinoceros who could not hear it because of the sound of the flowing waters of the river. Lord Krishna was annoyed. He disowned his mount and proceeded to Dwarka without it. The mount loved the environment of the green grass and stayed back. Hence the animal's presence in Kaziranga. It is also believed that the animal has no hair on its body as it was saddled for battle, according to a variant of the legend, and thus has hair only on its ears and tail tip![34] According to yet another legend, Lord Krishna's mount was readied for battle with armour. Later, it was considered unsuitable for war and abandoned; however the armour remained, thus explaining the folds of its armour-like skin which it bequeathed to its descendants.[35]

Over the past decades Assam and Bengal have seen several insurgent movements. This has impacted the protection of the rhinoceroses, to the extent that they were wiped out from Manas and had to be reintroduced from other protected areas. Its horn has made it a coveted target of poachers, which has always been a serious threat and accounts for the continuous decimation of rhino numbers.

The importance of the rhinoceros in the fauna of the Indian subcontinent has been recognized through its depiction on coins, stamps and commemorative issues. Despite the impediments and threats, the conservation efforts of the administration coupled with the local Assamese pride in their iconic symbol—a sentiment sadly lacking in Bengal—have played a prominent role in the rhino's successful survival and prosperity till date. At the same time such pride can and has led to some impediments in their reintroduction elsewhere, which is necessary for their long-term survival. We shall look at these issues in the final chapter.

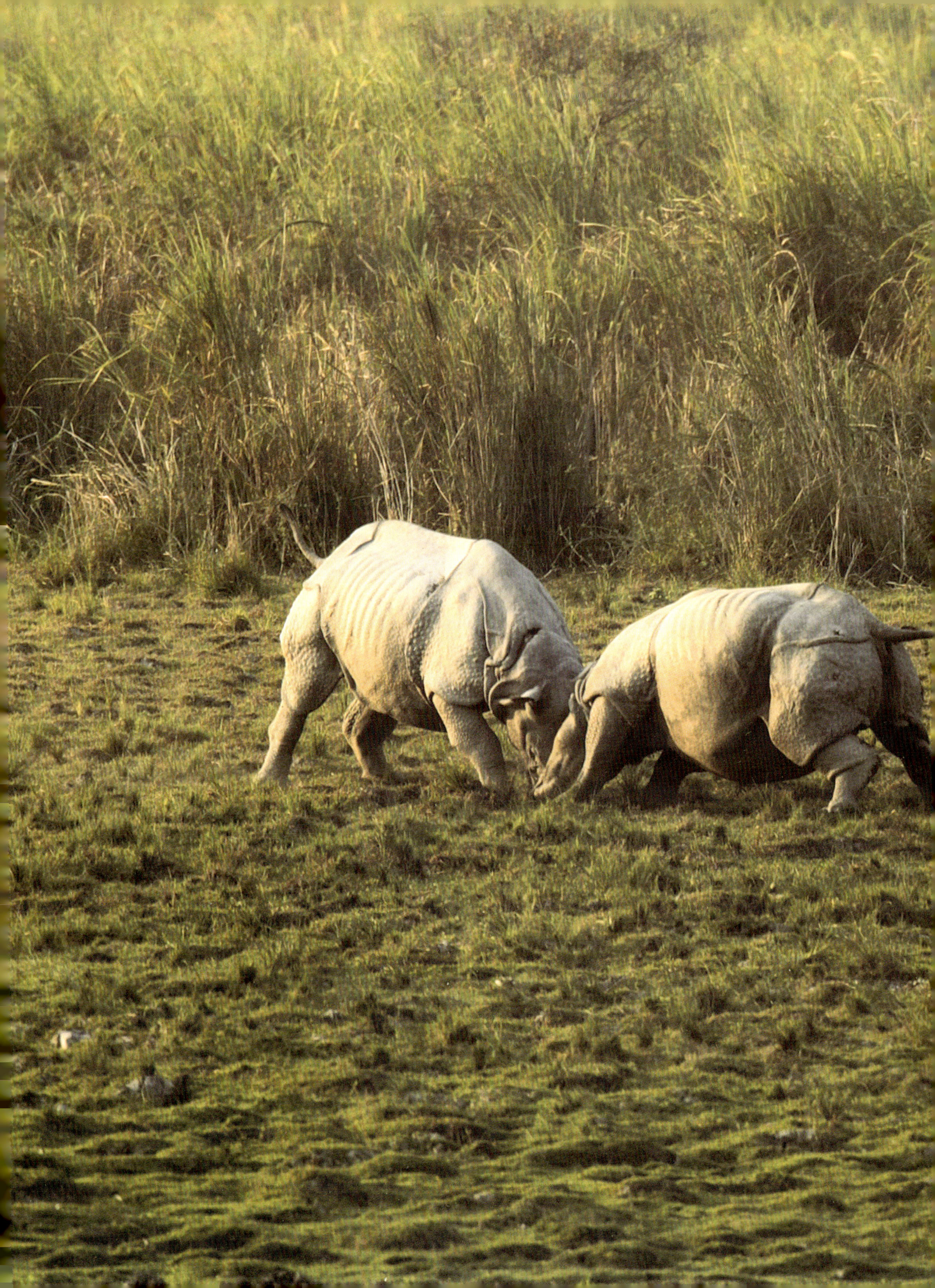

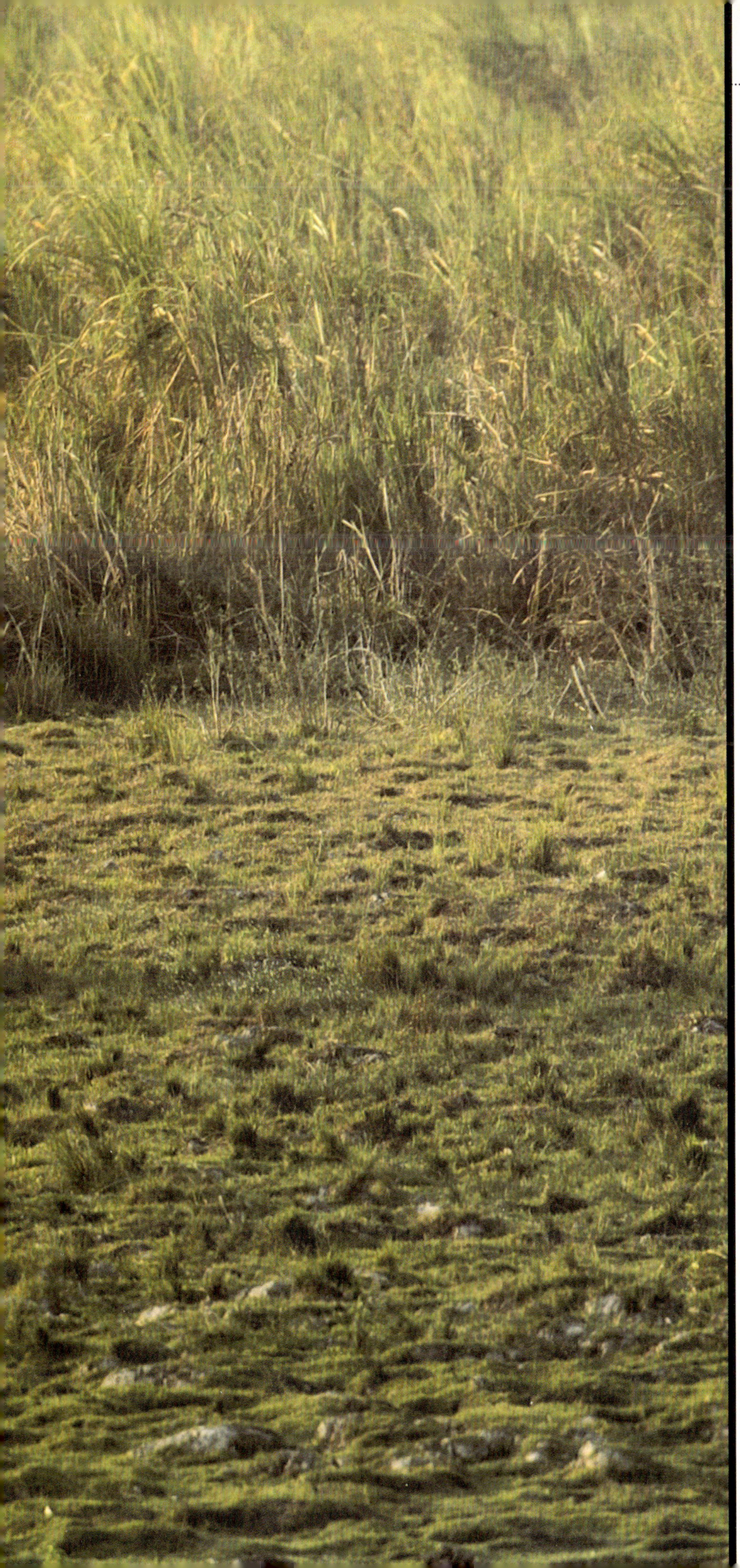

▪ Territorial fights among male rhinos can cause grievous injuries. Often bleeding animals have been spotted in Kaziranga.
Photograph: Kakubhai Kothari.

CHAPTER 7

Resurrection of the Unicorn

Divyabhanusinh

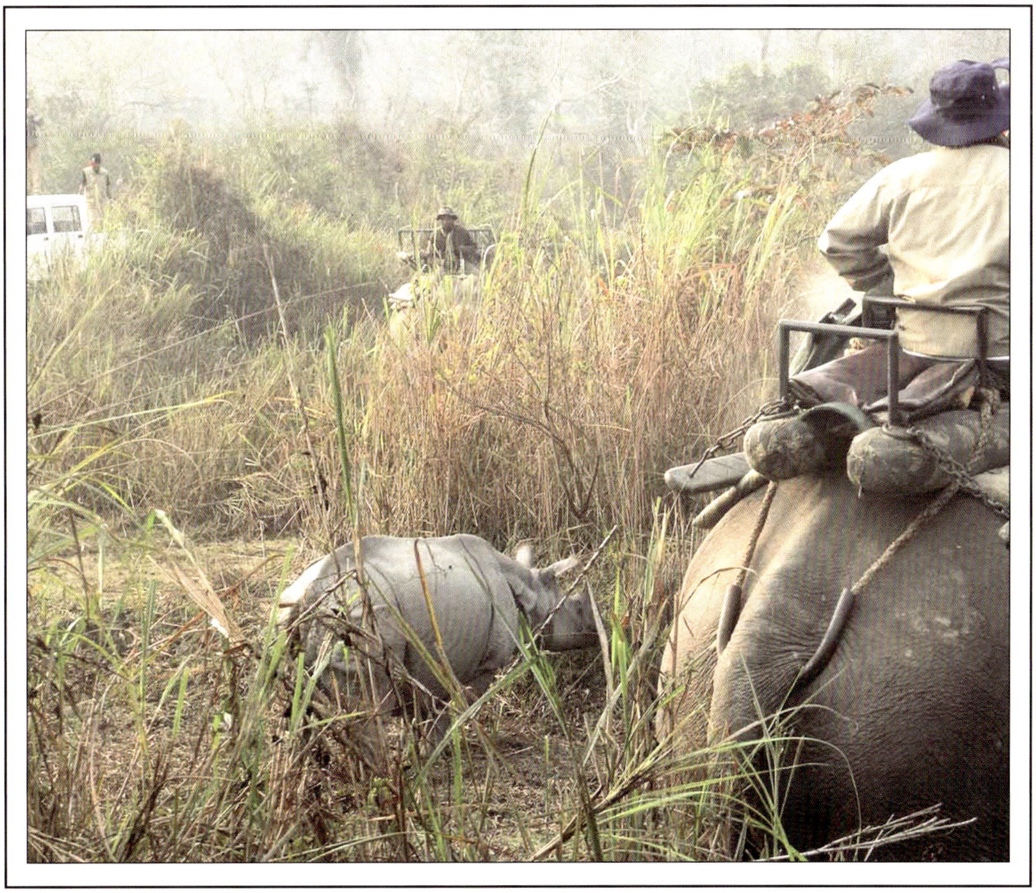

THE HUMAN QUEST for comfort from debility and disease through the use of plant and animal products is not new. Having tamed the environment, they made use of it to look after their own, particularly when they were ill. The political philosopher Thomas Hobbes (1588–1679) described the life of humans in nature as being "solitary, poor, nasty, brutish and short". While this is an extreme statement, he was not far off the mark when he called it short. To try and prolong their lives and enhance their physical power, humans often turned to animals that they saw as having qualities of strength, as a source of medicine.

The Rhino in Traditional Indian Medicine

Ancient Indians, as indeed humans elsewhere, observed and experimented with the fauna and flora which surrounded them. What followed was the creation of a system of natural medicine.

Charaka, "physician" of the 1st century CE, became an authority with his *Samhita* which described a system of medicine. In the 6th century Sushruta brought the extant knowledge up-to-date and thus was Ayurveda formalized as a system of medicine which is practised even today with developments incorporated over the ages. Without the benefit of scientific experiments which we take for granted today, the ancients including Sushruta categorized fauna to understand them better and to use their meats for medication for various ailments of humans.

Among the animals that lived on the banks of rivers, *kulachara*, were elephant, gaur, buffalo, barasingha, otters and the rhinoceros as noted in Chapter 2. We have seen that rhinoceros meat was recommended for various ailments and it was thus an important part of the pharmacopoeia of the time.

It will be noticed that in Ayurveda, it was the meat of rhinoceros that was consumed, while its horn had no specific use. It is of course possible, as is the case today, that the horn might have had some use along with the meat and other body parts in local and tribal practices which were not part of Sushruta's scheme of things. It is not out of place to record that the *Kama Sutra* of Vatsyayana Mallanaga, composed c. 300 CE, and which is arguably the most famous treatise on the art of conjugal and related pleasures, has detailed various types of love potions and aphrodisiacs recommended or in use at that time in ancient India. A careful examination of this vast array of medications, however, shows that the ancients did not use the rhinoceros horn or any part of the animal for amorous purposes.[1]

In our times, arguably the most authoritative survey of the use of rhinoceros body parts has been made by Esmond Bradly Martin. His extensive studies in India in the 1980s showed that, as in ancient times, rhinoceros horn is not used by Ayurveda practitioners. This system of medicine rarely included wild animal meats and body parts,[2] and thankfully even the ancient practice of using the meat of the animal appears to have been lost over the centuries.

However, another traditional system of medicine practised in India, the Unani system, did use the horn. This system is closely linked to Persian and Arabic traditional medicine which became prevalent with the advent of Islam in the Indian subcontinent. Unani is practised in numerous pockets of India today. In Mumbai and Gujarat, Martin observed rhinoceros horn products being recommended as an aphrodisiac by those practising this system of medicine. However, he found that the horn was not extensively used as in the past because of its high price, among other reasons.[3] Today, as a result of dwindling numbers of rhinoceroses worldwide

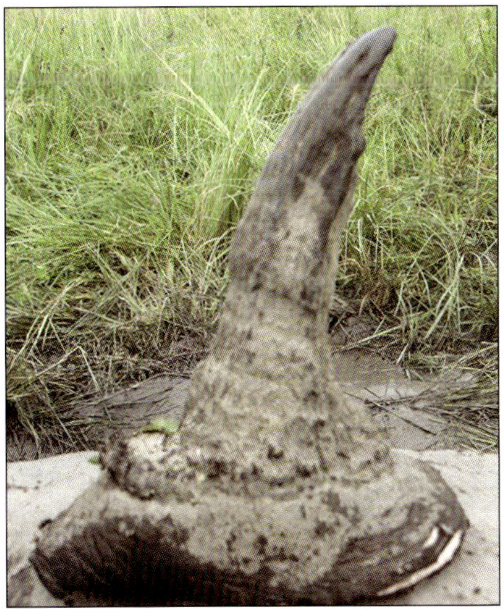

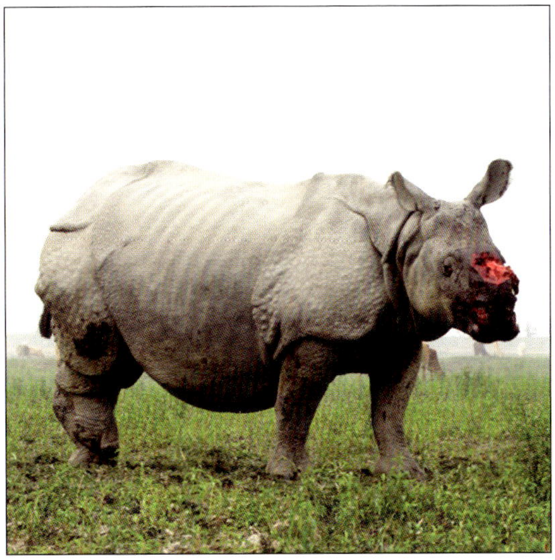

▪ Horn of despair: more costly than gold, this is the primary reason for rhinoceros poaching.
Courtesy Champak Deka.

▪ A wounded rhinoceros with its horn hacked off by poachers going through agonizing life before certain death, 2008.
Courtesy WCMP/Bibhav Talukdar, Aranyaka.

and stricter government controls, the open use and sale of the horn has come down considerably from what Martin saw in the 1980s, though illegal trade and poaching continue to be serious threats.

Even if the demand for rhino horn may have dwindled in India, this is not true of countries in East and Southeast Asia. Poachers continue to kill rhinoceroses in India for their horns because of the belief in their supposed medicinal value. It is not out of place to note here that poachers usually do not have the time to remove the skin or meat of the slain animal as they fear being caught by the authorities and coming to a sticky end—in Assam forest guards are authorized to shoot poachers if threatened by them. Since 2014, 260 poachers have been arrested in Kaziranga; of these 48 were arrested in 2017 and 12 were killed in the same year, according to a recent newspaper report.[4] The poachers thus would much rather make good their escape with the horn, which in any case is expensive enough to make their enterprise very profitable.

The Illegal Trade

This brings us to the problems of the illegal trade in the horn. At the outset let me state that rhinoceros horn has no medicinal properties other than the scientifically unsubstantiated belief that its use reduces fevers, as noted by Eric Dinerstein.[5] Even so, in earlier times it was supposed to have magical properties: some believed that if a poisoned drink was poured into a cup made from rhino horn, it would start frothing; others, that a horn placed under the bed of an expectant mother would assist her in childbirth; or that a horn left to soak in a bucket of water would turn it into an elixir of life to be sipped daily for its beneficial effects. Apart from the horn, the animal's urine is drunk by some persons for certain cures, and pieces of its hide and bone are worn as charms against sickness. Eating its meat is considered to be a passport to "the land of eternal bliss"![6]

We have seen that the Ayurvedic system did not use rhinoceros horn. Yet in the Chinese system of medicine and elsewhere in Southeast and East Asia, it is in great demand. In China, it is believed falsely to be a cure for several ailments. There is also a widespread impression that the Chinese use it as an aphrodisiac, which is untrue, though in some other ancient cultures it is an ingredient in "love potions". According to Richard Ellis, it is used for almost everything except as an aphrodisiac![7]

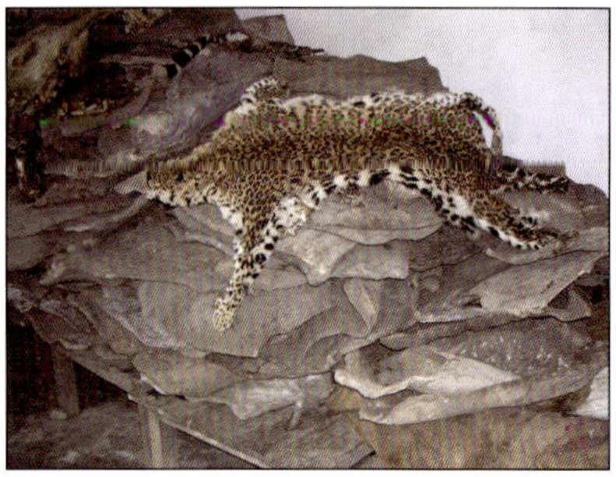

▪ Confiscated leopard skins and rhinoceros hides in Nepal.
Courtesy Forest Department, Assam.

In China the Indian rhinoceros horn is believed to be superior to the horns from elsewhere in Asia or Africa. As China sees greater prosperity of its population, the demand for the horn goes up. That apart, elsewhere in Asia where ethnic Chinese live and practise their system of medicine, the horn is in demand. In Indonesia the horn is used to treat high fevers by those who practise their own medicine system, while the people of Chinese descent there use it to treat typhus. The horn is sheared and boiled in water which is then drunk. In Malaysia, the horn water is sold and it is drunk for the treatment of many types of ailments such as fever, headache, toothache, earache, eye disease and thirst. In Myanmar, the horn is used for similar purposes by the Myanmarese and ethnic Chinese populations.[8]

The Japanese came in contact with the Chinese in the 6th century and became exposed to their culture and system of medicine. It may, therefore, be possible that they began using rhinoceros body parts in medicine from that time onwards. In Japan the horn is used to cure "*crysipelas* (an inflammatory disease which causes a bright redness of the skin usually on the face), weakness of the heart, diarrhea and vomiting".[9]

The illegal international trade in wildlife parts is serious business, being only lesser in terms of monetary value than the trade in narcotics. It is estimated to be worth US$ 20 billion in the global market.[10] The horn is an important component of this trade. India is one of the signatory countries to the Convention on International Trade in Endangered Species (CITES) who are clamping down on such trade. Consequently, the horn has become very costly—also because there are fewer and fewer rhinoceroses left worldwide to satisfy the demand. In spite of the best efforts of the government of Assam, where most rhinoceroses survive on the Indian subcontinent, poaching and trade continue. According to the figures of the state government, 193 animals were poached between 2001 and October 2014. Of these, Kaziranga National Park—arguably the animal's last stronghold—accounted for 126, whereas other protected areas of Orang accounted for 31, Pobitora 14, Manas 7, and other forest divisions accounted for 15 more. During the same period, 873 animals died of natural causes.

During the year 2013, 41 animals were lost to poaching whereas in 2014 up to November, 31 animals were poached in Assam. In Kaziranga the story is worrisome: between 1998 and 2006, 4 to 8 animals were poached annually. Unfortunately, between 2007 and 2010, and from 2012 to 2014, the number escalated to around 14–22 annually. In 2015 Kaziranga lost 17, in 2016 it lost 18, and 7 animals were poached in early 2018.[11] This is in addition to natural deaths caused by old age, disease or flooding which are quite substantial—85 animals were lost due to these reasons from November 1, 2015 to October 31, 2016. The temperamental Brahmaputra caused havoc with floods in 2017. Kaziranga, bound by this river in the north, was subjected to massive floods. According to a recent report, 361 wild animals lost their lives, of which 31 were rhinoceroses.[12] All this in spite of strict protection measures taken by the state government of Assam. In West Bengal and neighbouring Nepal too, the threat is very real. It is not out of place to note here that nearly 90 forest personnel and others have lost their lives in their line of duty in India and Nepal in recent times.[13] The figures speak for themselves, indicating that far greater efforts at protection are called for.

As we saw in the previous chapter, a primary cause of the intial decline of rhinoceros numbers was the sport of hunting. The body parts of animals that fell victim to shikaris—Indian potentates and British officers—were converted into a variety of artefacts to adorn their palaces and residences. With the enactment of the Wildlife Protection Act of 1972 all hunting finally came to an end.

India's Neighbours

Though this story of the unicorn is concerned with India alone, a political boundary is only a human concept. Our unicorn was also found in Bhutan where it may be extinct today, but it is still found in Nepal along India's borders. Here too, rhinoceroses were sought after as big game trophies. While forbidding shikar by outsiders without prior permission, the Ranas—the ruling class of Nepal—themselves organized hunts every year. In one such shoot in 1937–38 alone 38 rhinoceroses were hunted.[14] The opportunity to bag such trophies was also offered to visiting foreign dignitaries such as King Emperor George V of India, the Prince of Wales and other lesser beings (see Chapter 6, pages 107–08). During the visit of George V to India for his coronation as Emperor of India in 1911, he went to the Nepal Terai on a 10-day shikar trip in which 645 elephants were used. The shikar ended with a bag of 18 rhinoceroses, 39 tigers and 4 bears.[15]

There was also the ritual Tarpan ceremony which required the king to kill a male rhinoceros and "then make an offering of the animal's head and blood to the monarch's ancestors and pray for peace and prosperity of the Kingdom". This ceremony was required to be performed at least once in a king's reign. The last such Tarpan was performed by the late King Birendra Bir Bikram Shah.[16]

In spite of such depredations the monarchy and the ruling classes preserved the object of their sport for themselves in the tradition of princes elsewhere in the Indian subcontinent. As a result, according to Hemanta Misra, there were 1,000 rhinoceroses in Chitwan National Park alone in the early 20th century, and the population remained stable as the jungle and the animals were strictly protected.[17] However, Kees Rookmaaker paints a very gloomy picture of rhinoceros figures at the turn of the 20th century, as mentioned earlier. Be that as it may, it should be stressed again that the Ranas gave the animals protection which enabled them to hang on to existence.

Nepal experienced several political and social upheavals in the second half of the 20th century and the first decade of the 21st century. This had a direct impact on rhinoceros populations in the country. By the 1970s their numbers in all of Nepal had dwindled to between 90 and 108. With the accession of King Birendra Bir Bikram Shah in 1972 the animals were again strictly protected and their numbers bounced back to 450 in Chitwan alone. With the 2001 regicide, followed by the end of the monarchy and political turmoil caused by the communist insurgency, the forests once again became open to human depredation. In Chitwan alone where the population had grown to 550, it came down to 370 by 2005. In Bardia National Park 38 animals had been translocated in 1987 from Chitwan and another 46 had been translocated in the 1990s bringing their population to 89, which increased to 115. It appears that of these animals, only 15 survived the carnage by the year 2006. With stability in the country returning, the situation improved. In Chitwan, the number went up to 500, in Bardia, to 27, and to 7 in Sukla Phanta. The total for Nepal became 534 according to Hemanta Misra.[18] It may be recalled, however, that according to Eric Dinerstein, Chitwan had only 240 animals in 2011. Either way, the numbers are too small for comfort.

Habitat and Translocation

The most acute problem faced by all mammals, particularly the large ones such as lions, tigers, elephants, gaur, rhinoceroses and several others, is the ever-increasing pressure of exploitation by humans of the limited natural resources. This situation is worsening in India where we are 1.32 billion and growing. We want more and more space for our homes, farms and factories. This alone has fragmented the habitat of all mammals including the rhinoceros. A species that once thrived from the Indus and Punjab in the west to the Myanmar border in the east along the foothills of the great mountain ranges of the Himalaya, today is making its last stand in a few isolated pockets. The table shows their fragmented populations in 2015.[19]

The total world population of greater one-horned rhinoceroses in the wild today is 3,500-plus, a distinct improvement on the figure of about 200 estimated in 1900, as noted earlier. In spite of this, a grim picture stares us in the face.

Kaziranga is the last stronghold with Chitwan, Jaldapara and Orang way behind with three-digit figures. The populations in the rest are confined to two digits, while one protected area has only eight animals. In all but Kaziranga and possibly Chitwan the numbers are indeed small, and their habitats are fragmented, thus the animals are isolated. This results in inbreeding with long-term consequences for their survival which are not fully known. Besides, an outbreak of disease can severely impact these relict populations. This situation calls for scientific management interventions from the governments concerned. In this respect, no single effort can be considered adequate, it must be a continuing process and not a one-time intervention.

Two animals, the Asiatic lion and the greater one-horned rhinoceros, one from the far west and the other from the distant east—in Gujarat and Assam respectively—have surprisingly similar stories to narrate. It was noted by one authority—but denied by several others—that there were only about 12 lions left in the Gir forest at the turn of the 20th century. Lord Curzon, Viceroy and Governor-General of India, cancelled his proposed shikar of a lion during his state visit to Junagadh in 1900. He encouraged Nawab Rasulkhanji (1892–1911) to continue his efforts which were already underway to protect lions in his domains. As noted earlier, tradition has it that the Vicereine Lady Mary Curzon went to Kaziranga to see rhinoceroses but had to be satisfied with seeing only the animal's spoor. It was believed that only 12 rhinoceroses survived at the time in the Kaziranga area.[20]

Equally interesting is the fact that these two majestic animals are considered to be the unique possessions of the Gujarati and Assamese people respectively. Gujarat, no doubt, claims the lion as its own since it does not exist elsewhere. On the other hand, the rhinoceros does exist outside of Assam. It is acknowledged that regional pride has helped the animals to survive in their respective regions to a point. However, there is a downside too.

One of the effective ways to save a species is to create new homelands in its former range by translocation and subsequent protection to the animals in their new habitats. Towards this end the government of Saurashtra state had allowed lions to be translocated to Chakia forest near Varanasi in Uttar Pradesh in 1957. There they prospered but ultimately failed to survive due to poor management. In the case of the rhinoceros, animals were translocated from Assam to Dudhwa National Park in Uttar Pradesh. There the story has a happier ending to which we shall come shortly. Both these translocations were made in what now appears to be a distant open-minded past. Now, Gujarat does not wish to part with a few lions to Kuno Palpur in

TABLE: RHINOCEROS NUMBERS IN 2015

Protected area	Area in sq. km	No. of Rhinoceroses
India		
Kaziranga NP	884.43	2,401
Jaldapara NP	216	200
Orang NP	78.8	100
Pobitora WLS	38.81	92
Gorumara NP	79	56
Dudhwa NP	660	32
Manas NP	500	32
Nepal		
Chitwan NP	932	605
Bardia NP	968	29
Sukla Phanta WR	305	8

Note: NP= National Park, WLS= Wildlife Sanctuary, WR = Wildlife Reserve

Madhya Pradesh and Assam is reluctant to relocate rhinoceroses anywhere outside the state. Thus regional pride acts as a brake to management initiatives for a long-term survival plan for these species which are in a vulnerable position, despite the fact that in both cases the numbers are rising or are steady overall in their current habitats in the two states.

The translocation of rhinoceroses in Nepal seems to be a success story, as they appear to be holding on and increasing at least in one case, in spite of the setbacks that they suffered due to political and social upheavals in the country. In India too, the experiment has met with success till date, which needs to be carried forward in the future. In 1979, the Indian Board for Wildlife (now known as the National Board for Wildlife) set up a committee to evaluate the possibilities of finding a new home for the rhino. After extensive studies, it was decided to go ahead and translocate rhinoceroses from Assam to Dudhwa National Park. The saga of this effort of moving rhinoceroses from Assam to Uttar Pradesh by road and air and their subsequent release has been meticulously recorded by Samar Singh and Kishor Rao, Joint Secretary and Deputy Director, Ministry of Environment, Government of India. Detailed planning and efficient execution ultimately paid off.[21]

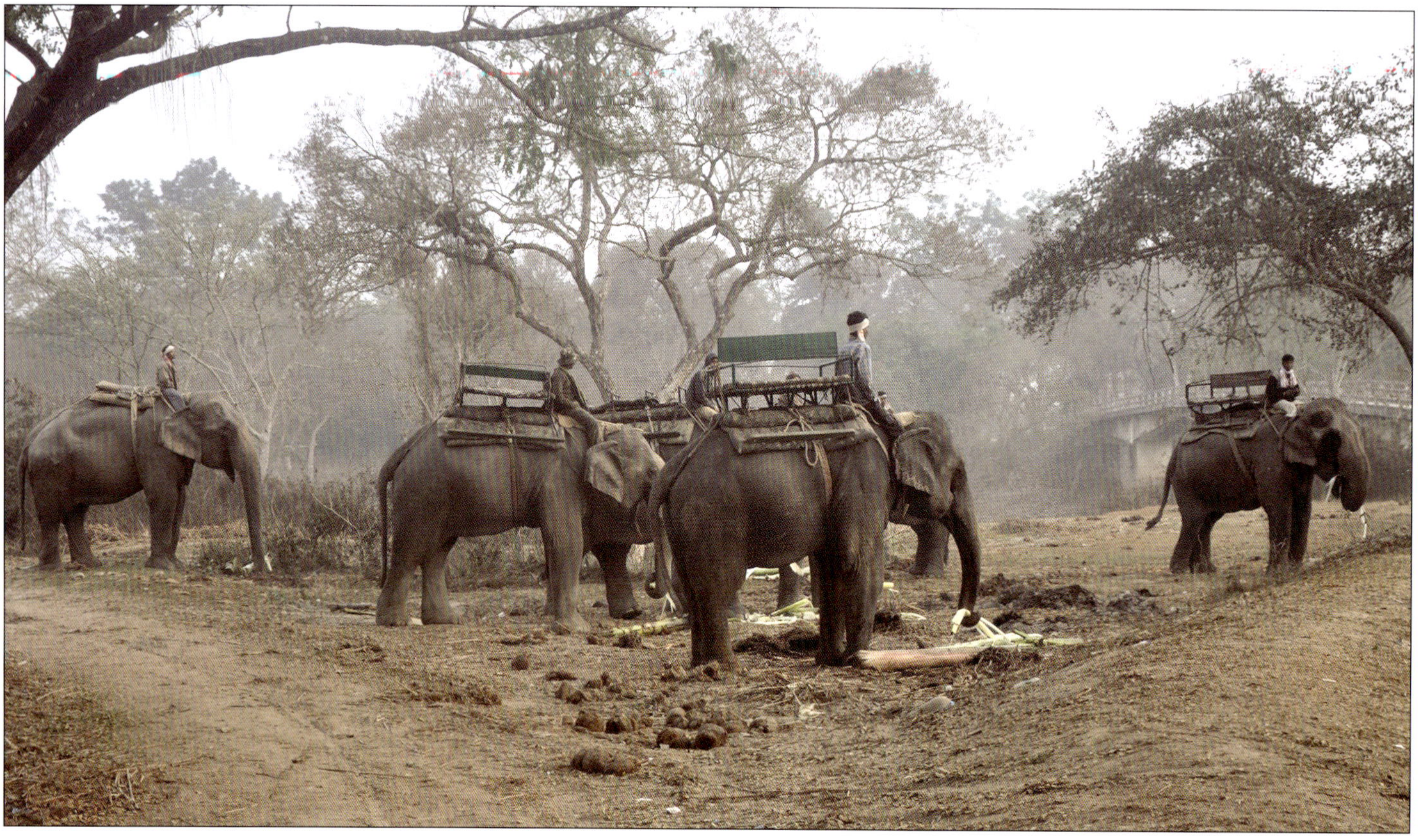

Pages 124–29
Translocating rhinoceroses from Kaziranga National Park to Manas National Park, Assam, 2012.

▪ Elephants being readied for the capture operation.
Courtesy P. Das, Forest Department, Assam.

Dudhwa has an area of over 600 sq. km, in which a large enclosure of 27 sq. km was made for the translocated animals. In 1984, five rhinoceroses, one sub-adult and two elderly females, one young adult male and one older male captured from Pobitora Wildlife Sanctuary in Assam were transported by air and road to their new location.

Unfortunately, neither of the old females survived after translocation. In the subsequent year, the Nepalese government generously gave four females from Chitwan National Park. At the present, three animals survive from the founding stock.[22] This programme of the government is being assisted by the World Wide Fund for Nature–India (WWF–India) which has among other things evolved a unique ocular protocol to identify each animal. It is heartening to record that the population has grown to 23 today[23] though this must be treated only as a first step.

Subsequent to this translocation the Assamese people's pride in their *gainda* became apparent. Now the government of the state does not wish to part with any more animals for translocation outside it. As such, the most promising supply reservoir of animals to create new homes has dried up, though rhinoceroses are still sent from one protected area to another within Assam.

The Assam government undertook an ambitious programme of sending rhinoceroses to Manas National Park where once they had been found in abundance but had become locally extinct because of political unrest caused by insurgency. Between 2008 and 2012 Pobitora sent 10 animals including 1 male and 3 female adults, 2 male and 2 female sub-adults, and 1 male and 1 female calves. In 2012 Kaziranga sent 4 adult females, and 3 male and 1 female calves. So a total of 18 rhinos were translocated in four years' time. Today the population is at 32 animals, though a couple of rhinoceroses have fallen prey to poachers.[24]

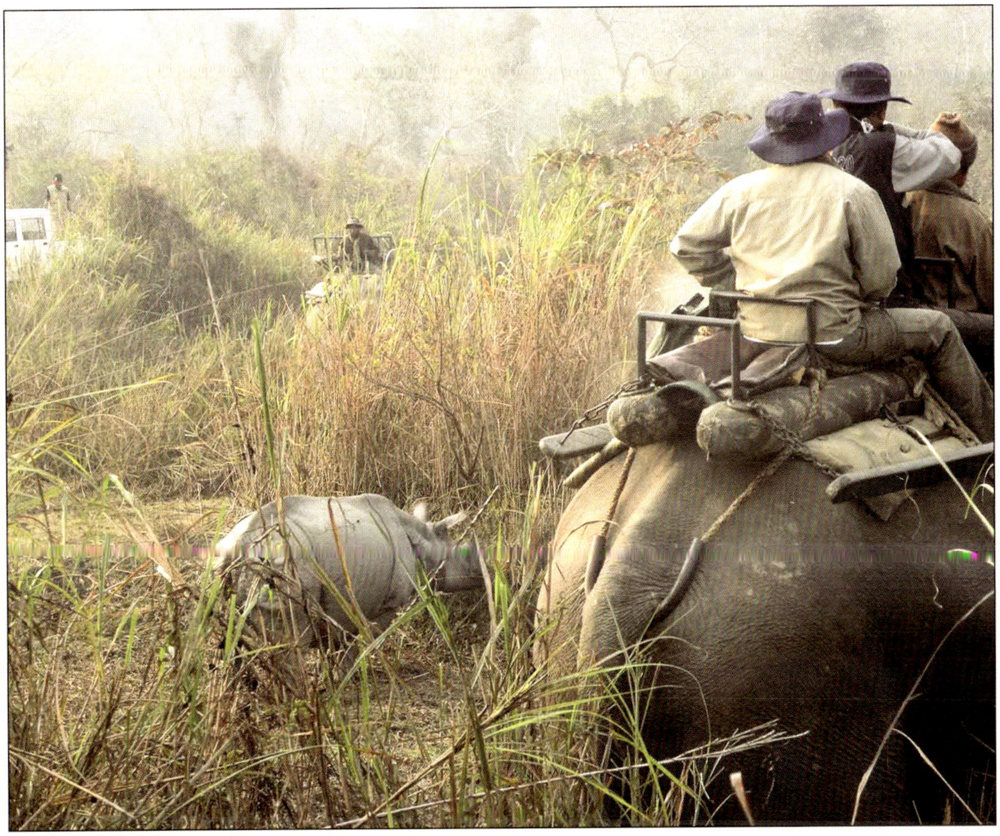

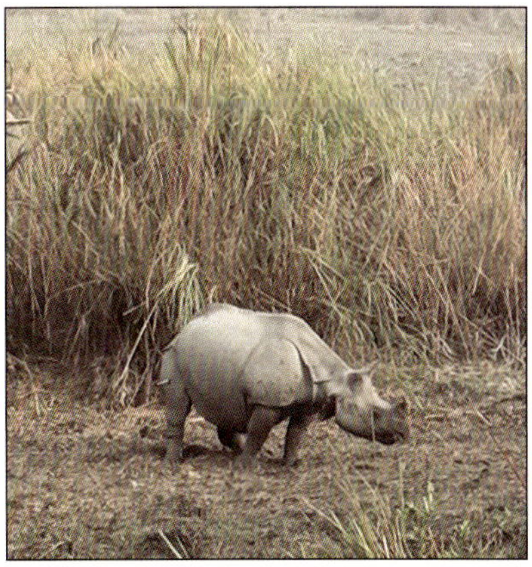

▪ Rhinoceros being tranquillized by a dart gun.
Photograph: Deba Dutta, WWF–India.

▪ A darted animal with the tranquillizing dart in its right rear leg.
Courtesy Amit Sharma, WWF–India.

▪ Preparing to transport the animal.
Courtesy Amit Shah, WWF–India.

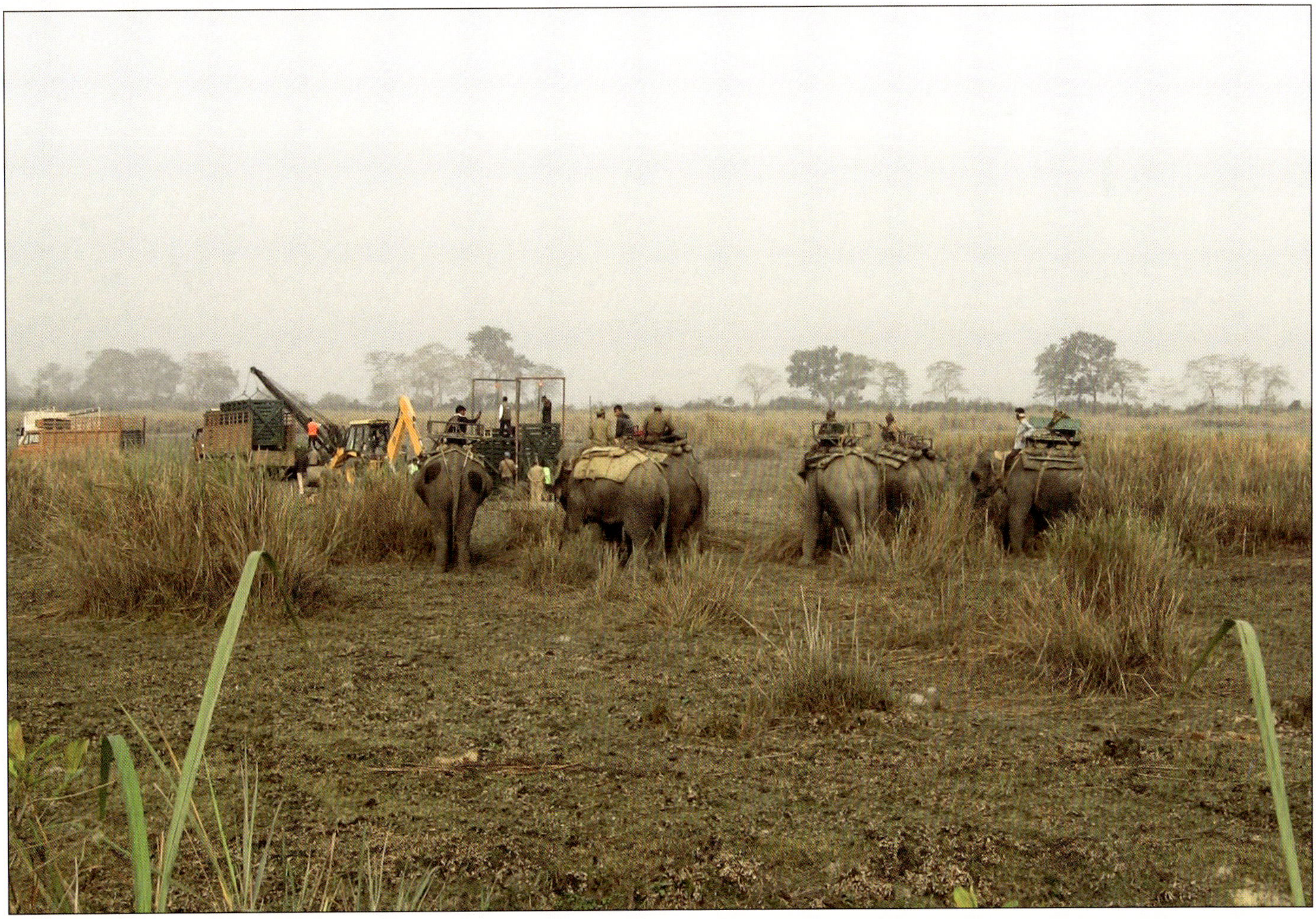

■ Darted animals undergo veterinary examination and are radio collared.
Courtesy Amit Shah, WWF-India; P. Das, Forest Department, Assam.

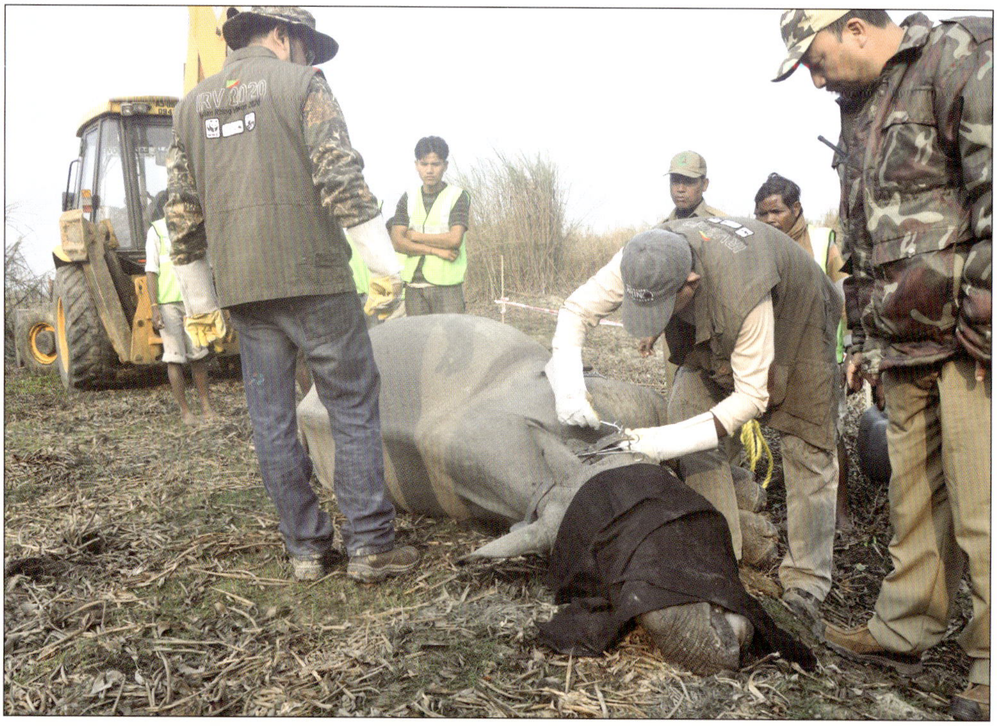

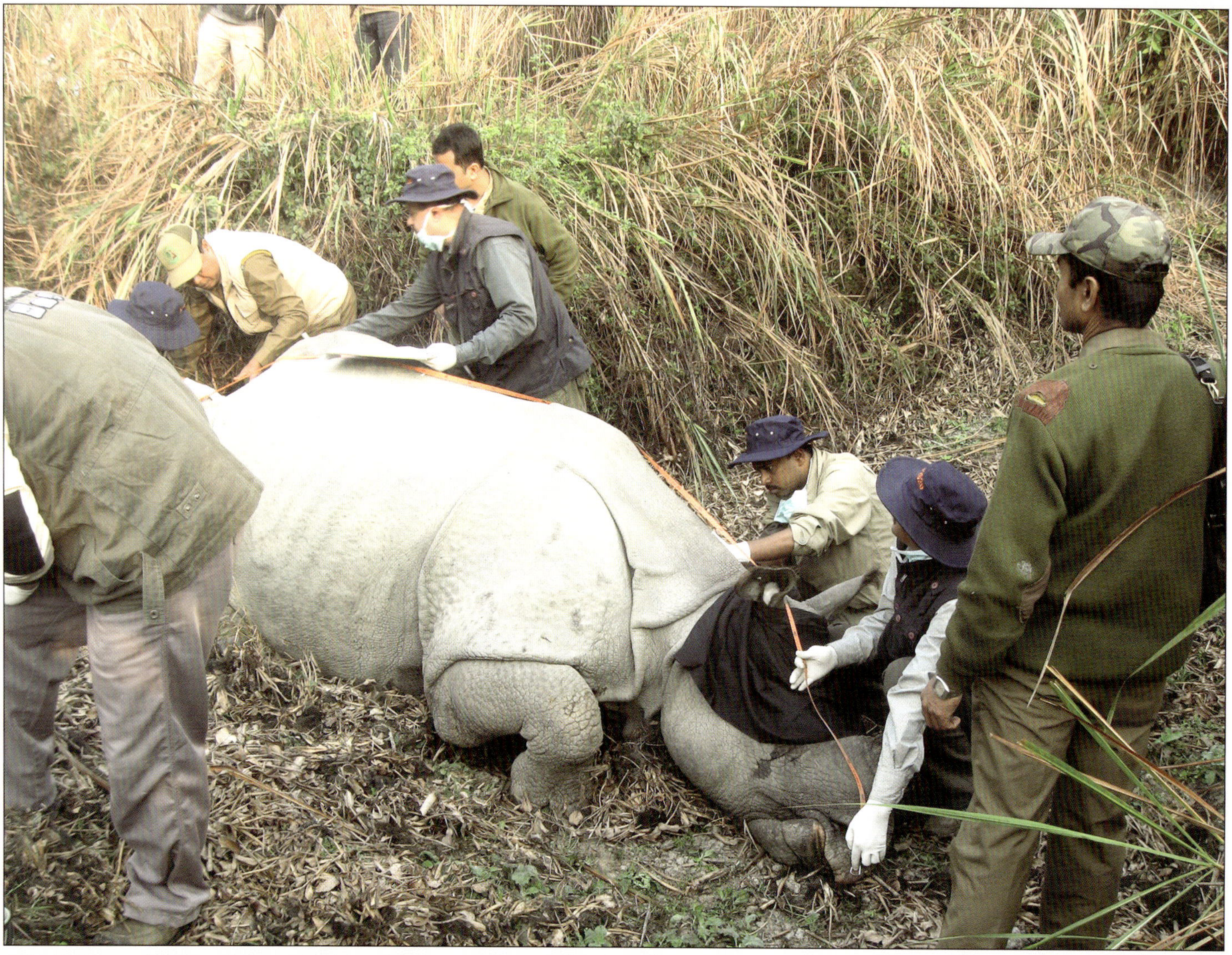

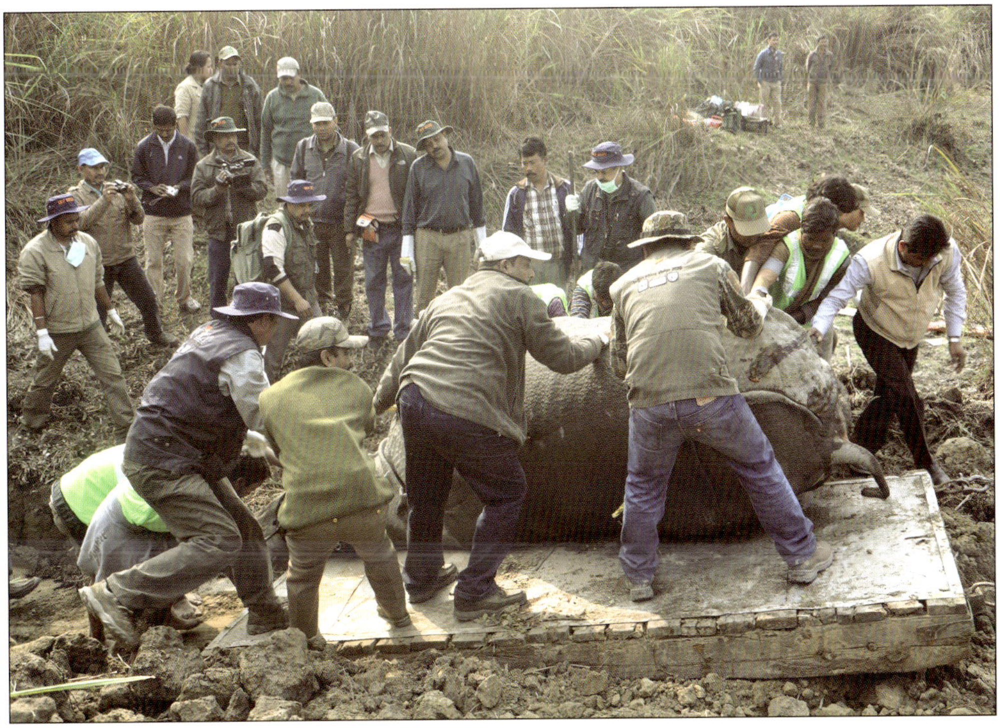

■ The animal is rolled on to a sledge and slid into a truck.

Courtesy P. Das, Forest Department, Assam.

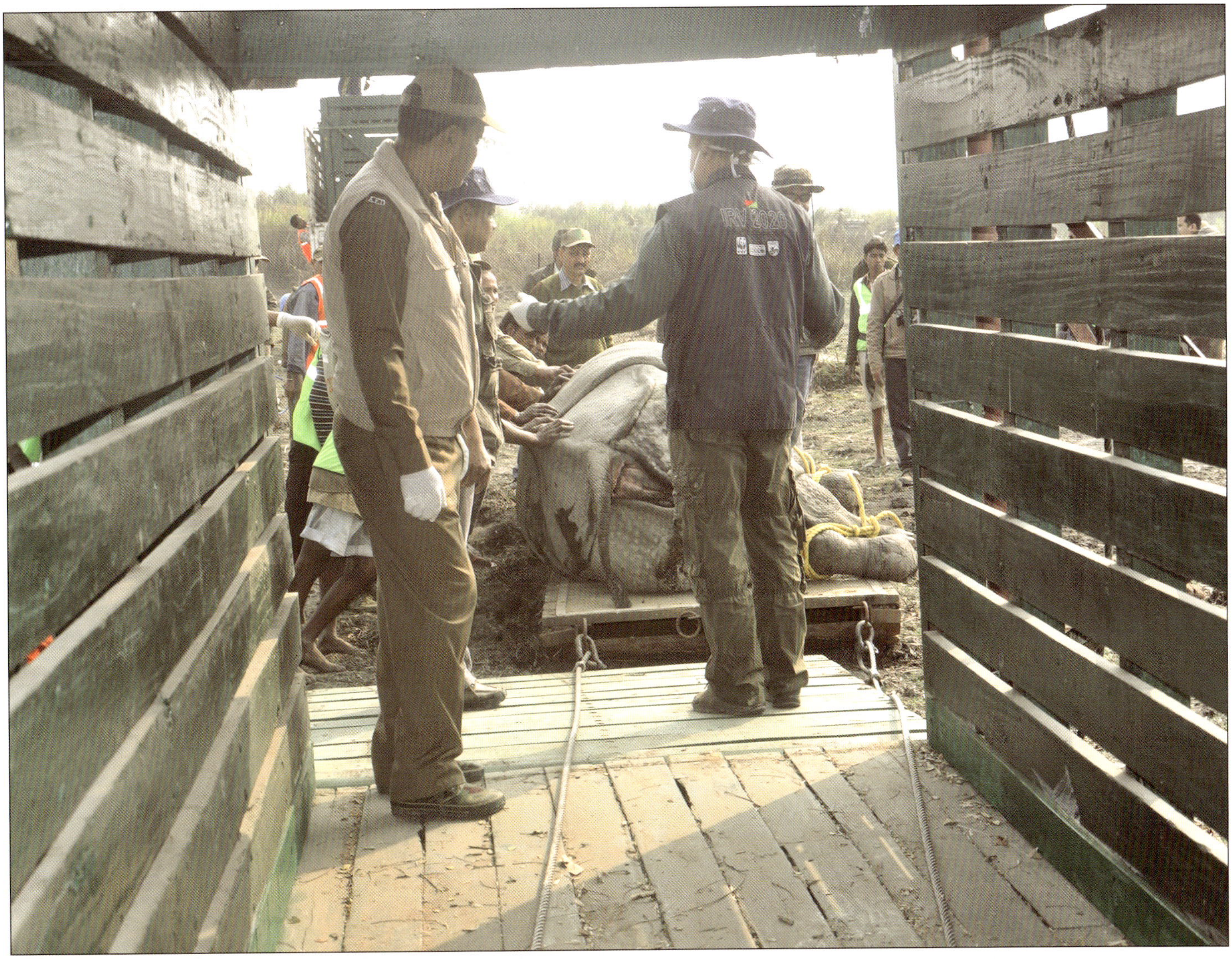

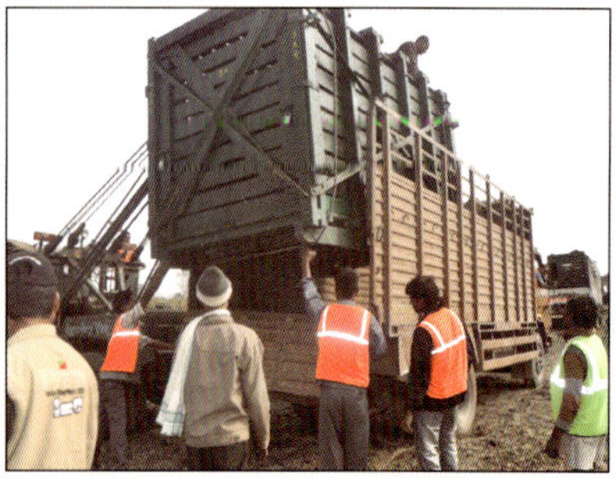

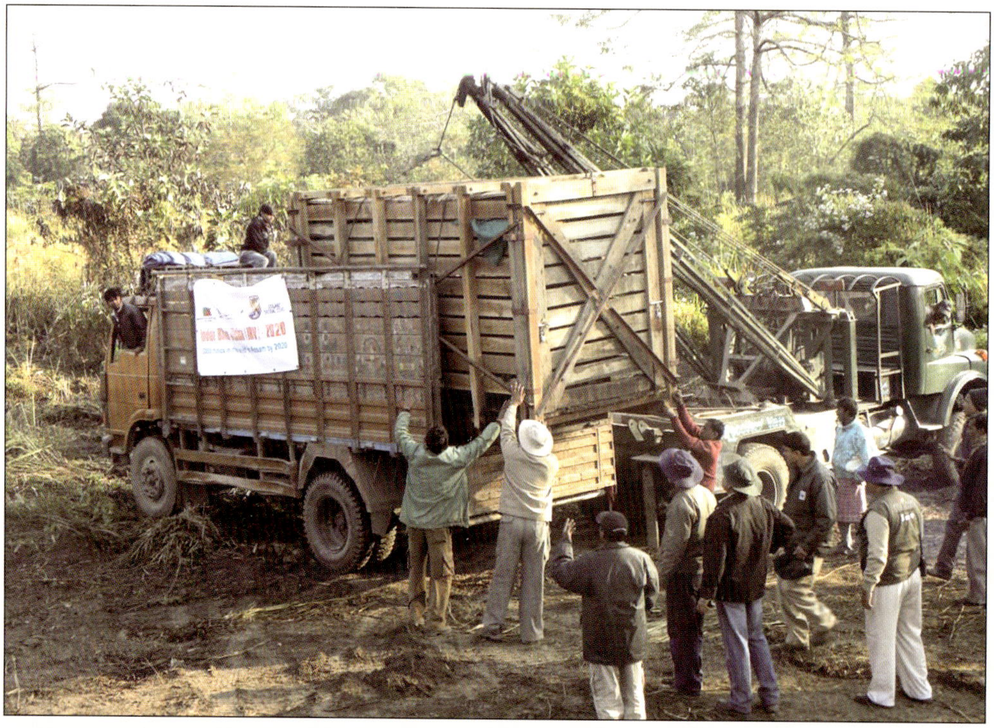

- The animal departs from Kaziranga National Park.
  Courtesy P. Das, Forest Department, Assam.

- Unloading at Manas National Park.
  Courtesy Deba Dutta, WWF-India.

- Freedom at last for two rhinoceroses after a long, gruelling translocation experience.
  Courtesy Iusuf Khan, WWF-India.

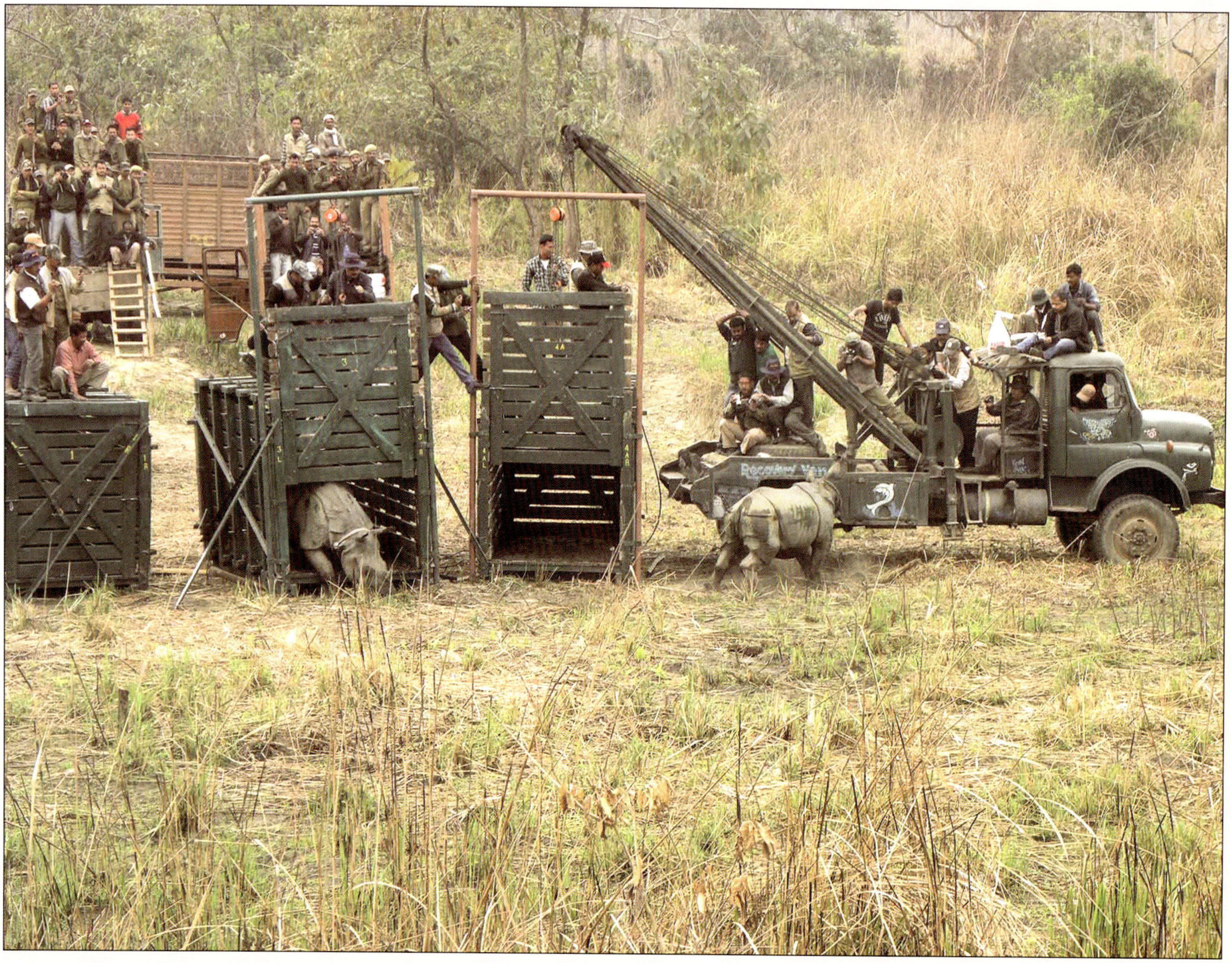

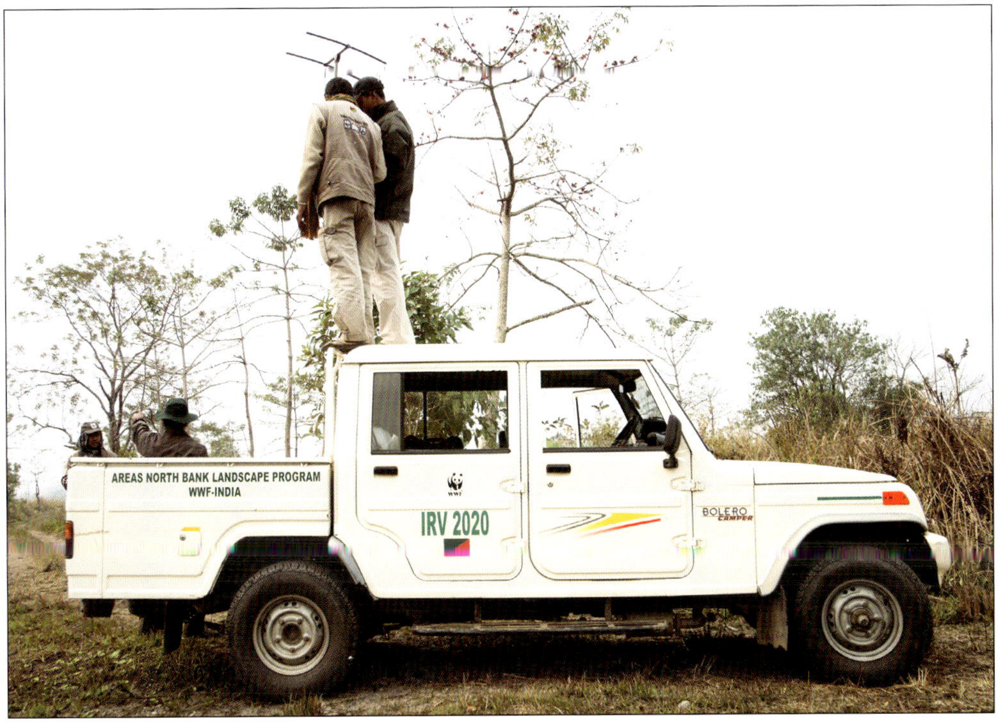

■ Once the rhinos are released after being radio collared, they are monitored continuously by the forest department officials. This is essential to ensure their wellbeing.
Courtesy Anil Cherukupalli, WWF-India, 2011.

Major threats to the rhino population are annual flooding of the Brahmaputra river, road kills as well as the possibility of an epidemic which could drastically affect their fragmented populations. Above all, there is the continuous and ever-present poaching that goes on as the horn pays rich dividends. The figures of poached rhinoceroses quoted above speak for themselves. This pressure can only increase with growing demand for the horn over large parts of Asia. When the population of a species reduces to such an extent that it becomes necessary to count its individuals, it is a warning bell loud and clear, for all to hear. The poor rhinoceros has suffered and continues to suffer for its horn. This horn of despair is a monument to the greed, folly and ignorance of human beings, for it cannot do for them what they believe it can.

Still, any way one looks at the condition of the unicorn, its conservation has been a success story, in spite of all the threats. Its resurrection from supposedly less than 200 in 1900 to more than 3,500 in 2015 is impressive. Assam is home to 70 per cent of this population. Plans are afoot to reintroduce the rhino in Dibru Saikhowa National Park and Laokhowa and Burachapori Wildlife Sanctuaries in the state. With these and other measures, it is hoped that the upward swing will continue and rhinoceros numbers will increase to 4,000 by 2020.[25]

Appendix

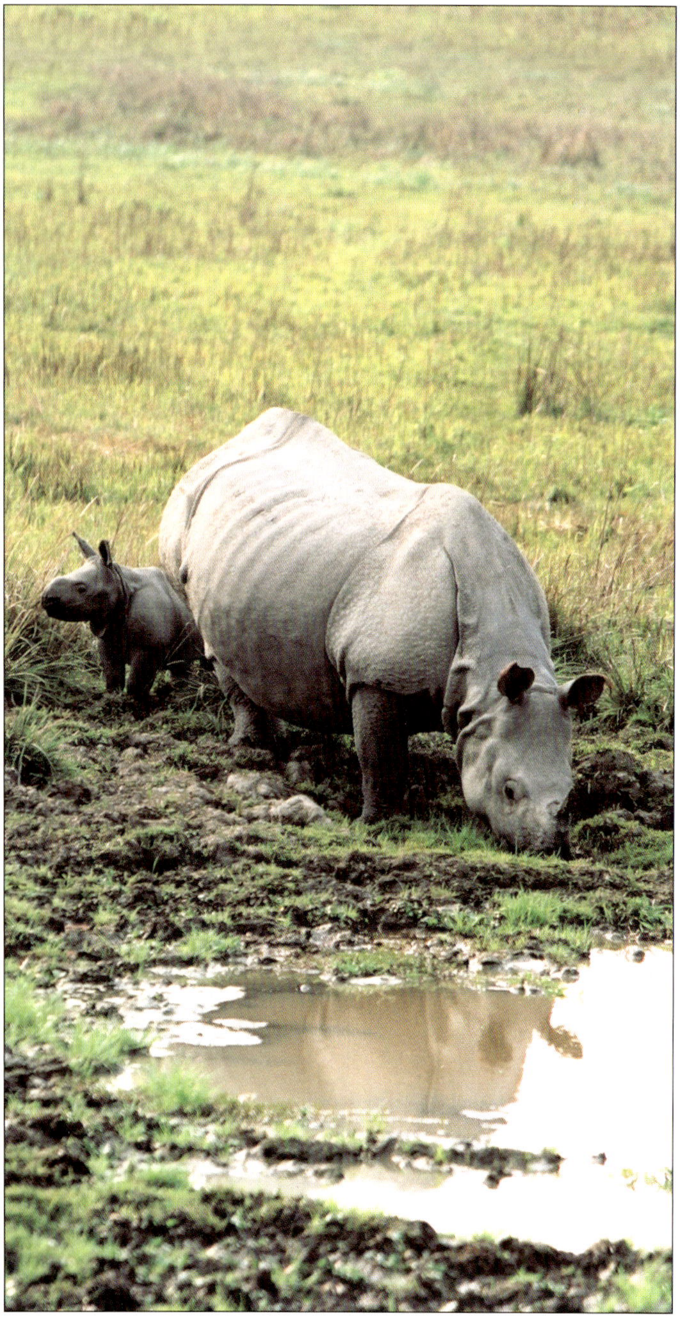

List of Fauna

Bear, Brown	*Ursus arctos*
Bison, American	*Bos bison bison*
Buffalo, Wild	*Bubalis arnee*
Cheetah, Asiatic	*Acinonyx jubatus venaticus*
Crocodile, Estuarine/Saltwater	*Crocodilus porosus*
Deer, Brow-Antlered/Thamin	*Rucervus eldii thamin*
Deer, Hog	*Axis porcinus*
Elephant, Asian	*Elephas maximus*
Gaur	*Bos gaurus*
Lion, Asiatic	*Panthera leo persica*
Peafowl, Indian	*Pavo cristatus*
Rhinoceros, Greater One-horned	*Rhinoceros unicornis*
Rhinoceros, Javan	*Rhinoceros sondicus*
Rhinoceros, Sumatran[1]	*Dicerorhinus sumatrensis*
Sumatra and peninsular Malaysia	*D.s. sumatrusis*
Borneo	*D.s. harrissoni*
Southeast Asia (could be extinct)	*D.s. lastiotis*
Rhinoceros, Black[2]	*Diceros bicornis*
Namibia, W and SE Africa	*D.b. bicorni*
Historically in Nigeria, Cameroon, Chad, Sudan and Central African Republic	*D.b. longipes*
Kenya, Rwanda (could be extinct) and N. Tanzania, historically also in Sudan, Ethiopia and Uganda	*D.b. unichaeli*
Rhinoceros, White[3]	*Ceratotherium simum*
Tiger	*Panthera tigris*
Wild Boar	*Sus scrofa*
Wolf, Hokkaido	*Canis lupus hattai*
Wolf, Japanese[4]	*Canis lupus hodophilax*

NOTES

1 Dinerstein, 2011, p. 180.

2 Ibid., p. 177.

3 Ibid.

4 Walker, 2005, pp. 39–42.

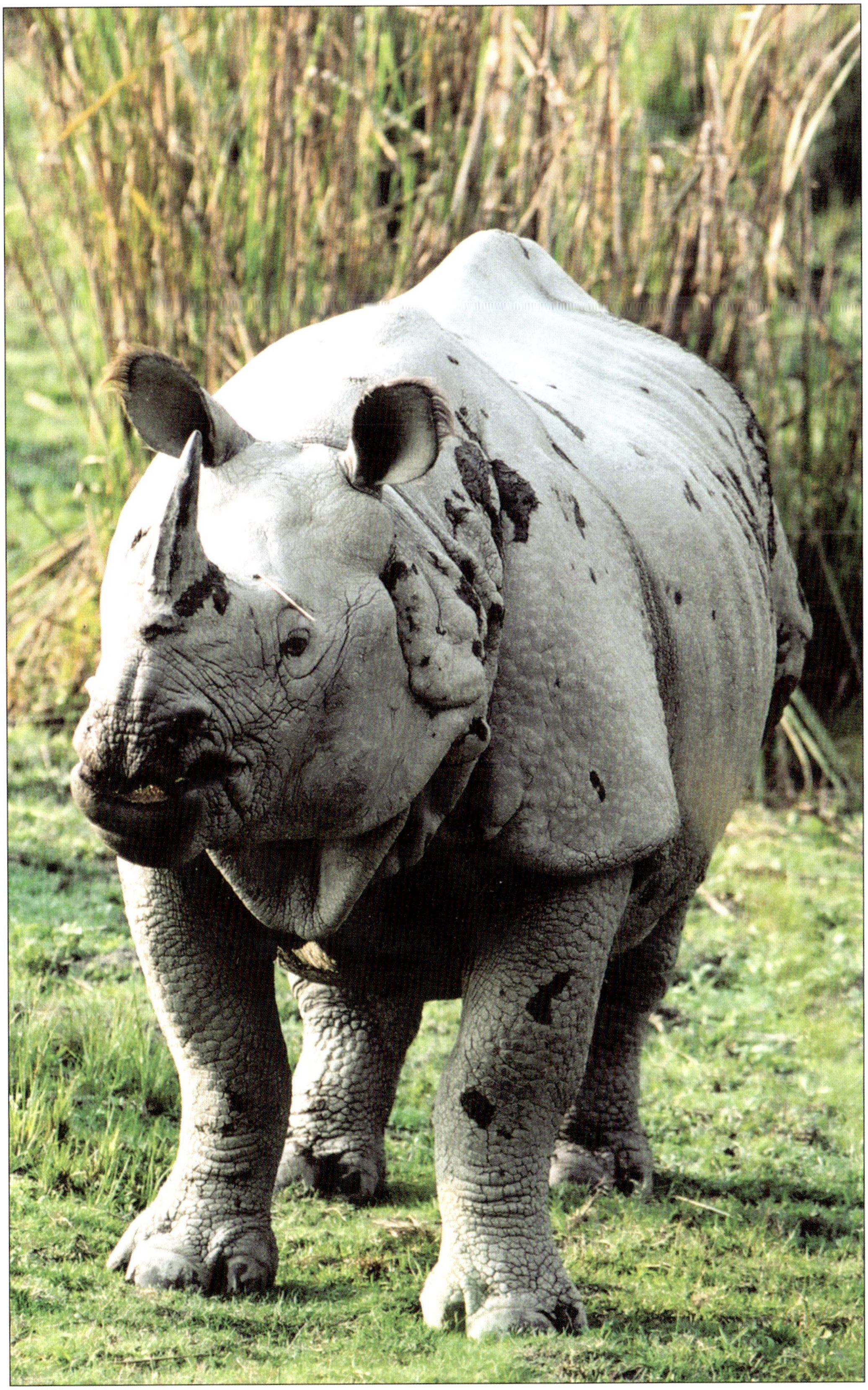

■ The rhino prefers to wallow and is often found near waterbodies where it also finds green grass. Kaziranga.
Photograph: Kakubhai Kothari.

■ The rhino appears to be at peace with itself in Kaziranga.
Photograph: Kakubhai Kothari.

Maps

Historical distribution of Asia's three species of rhinoceros

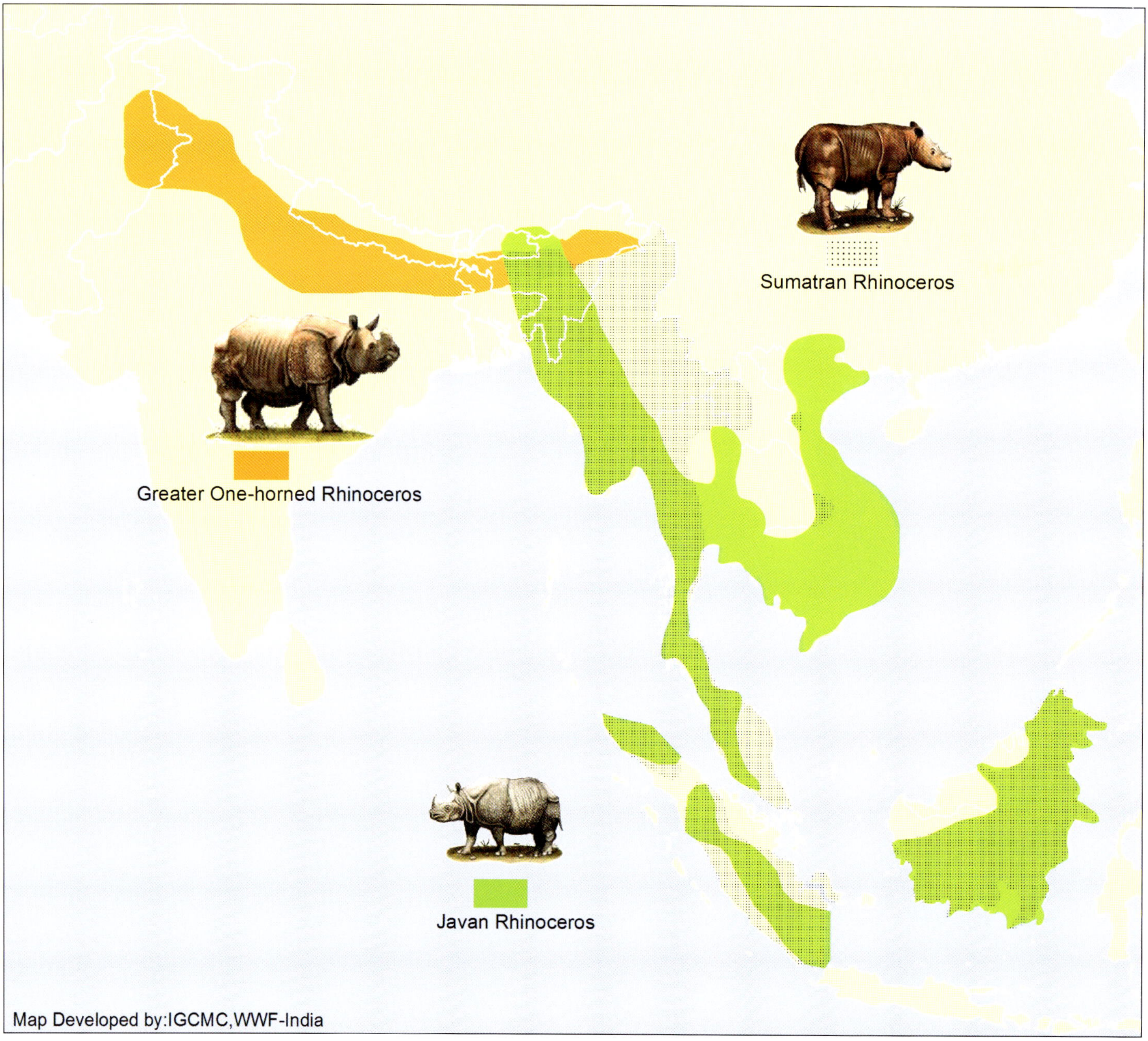

Courtesy WWF–India, 2017.

Historical and current distribution of the Greater One-horned Rhinoceros

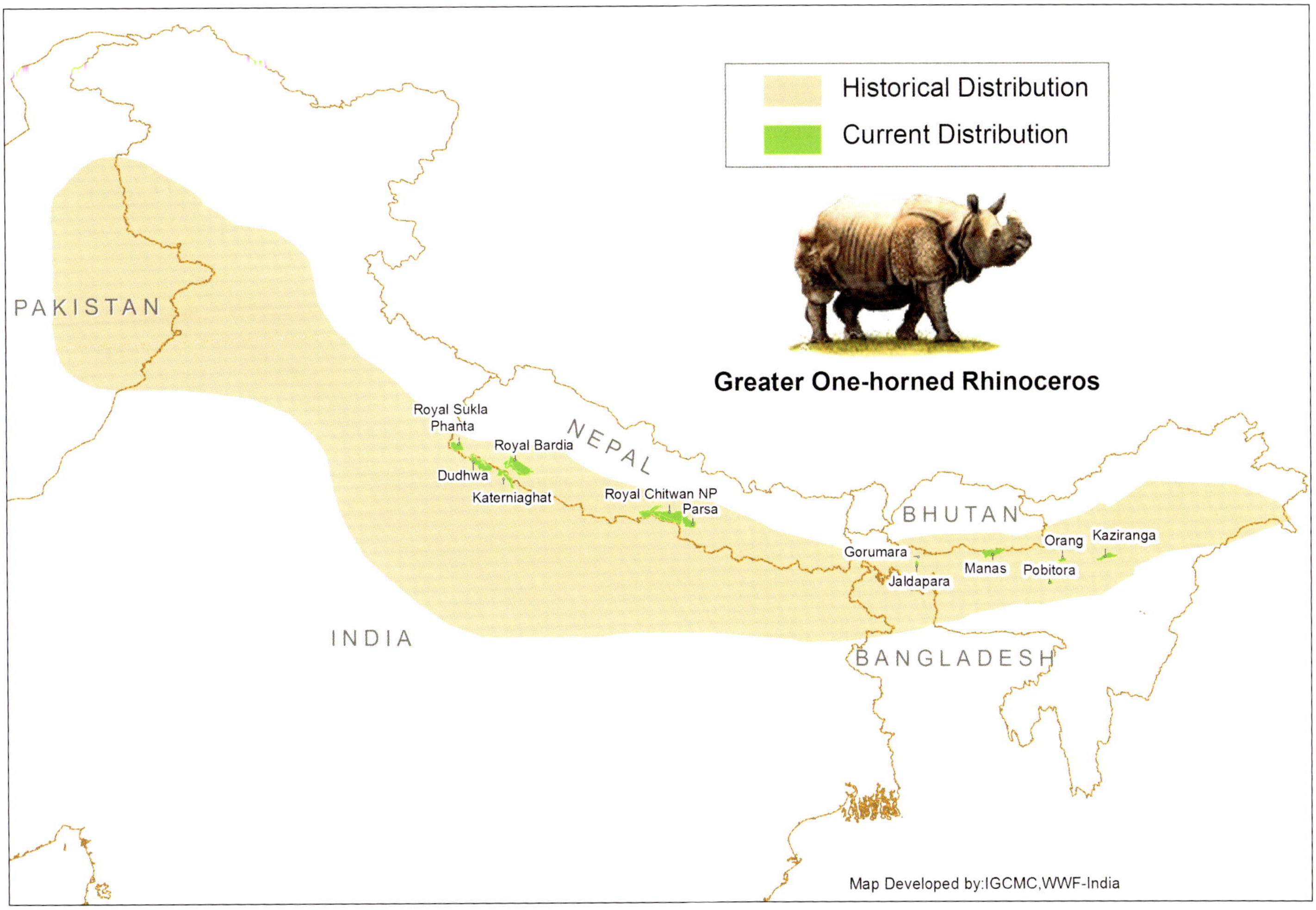

Courtesy WWF–India, 2017.

CHAPTER 1 (pp. 12–29)

1 Allaby, 1985, p. 482.
2 Dinerstein, 2003, pp. 10–24; 2011; 2015.
3 Jerdon, 1874, p. 234.
4 Sanyal, 1892, p. 131.
5 Jerdon, 1874, p. 234; Sanyal, 1892, pp. 131–32.
6 Sterndale, 1884, pp. 410–12.
7 Lydekker, 1924, pp. 34–40.
8 Prater, 1948, pp. 193–94.
9 Rookmaaker, 1997.
10 Dinerstein, 2011, pp. 179–80.
11 Rookmaaker et al., 2016, pp. 7, 16.
12 Monier-Williams, 1985, pp. 690, 799; 1988, p. 335; Apte, 1992, p. 390.
13 Dave and Chandra, 1953, species nos. 334, 335 and 336.
14 Gotch, 1995, p. 625.
15 Ibid.
16 OED, 1977, Vol. II, p. 3512.
17 Thompson, 1956, p. 499b.
18 Dinerstein, 2011, p. 179.
19 Ibid., p. 148.
20 Smith et al., 1989, p. 703.
21 Pastoureau, 2011; Walker, 2005; Isenberg, 2001.
22 Vogel, 1962; Nair, 1977.
23 Divyabhanusinh, 2006; 2008; Sukumar, 2011.
24 Sachau, 1983, pp. 204–05.
25 Gibb, 1971, p. 596.
26 K.K. Basu, 1944, pp. 145–46; Professor Chetan Singh, personal communication.
27 Ridley, 2004, p. XV.
28 Clarke, 1986.
29 Mackay, 1997, p. 94.

CHAPTER 2 (pp. 32–43)

1 Badam and Jayakaran, 1993, p. 249.
2 Laurie, 1978, p. 6; Laurie et al., 1983, p. 2.
3 Rookmaaker, 1984; 1999; 2000; 2002.
4 G.L. Badam, personal communication.
5 Badam and Jayakaran, 1993, pp. 244, 259.
6 Cockburn, 1883, pp. 56–64.
7 Blumenschine and Chattopadhyaya, 1983, p. 283.
8 Badam, 1979.
9 Wilson and Mittermeier, 2011, p. 179.
10 Pal, 2002, p. 61.
11 Chattopadhyaya, 1996, p. 464.
12 Chattopadhyaya, 1999, p. 128.
13 Lukacs et al., 1996, p. 303.
14 Thomas et al., 1995b, pp. 31–32; 1996a, pp. 257–58.
15 *Indian Archaeology 1971–72–A Review*, p. 49.
16 Misra et al., 2003, p. 24.
17 Joglekar et al., 2003, p. 75.
18 Ibid., p. 109.
19 Ibid., pp. 109, 115–16.
20 Momin et al., 1973, p. 801.
21 Ibid.
22 Shah, 1980, p. 70.
23 Clutton-Brock, 1965, pp. 9–10.
24 Zeuner, 1952, p. 129.
25 Clutton-Brock, 1965, p. 37.
26 Divyabhanusinh, personal communication.
27 Wakankar, 1973.
28 Varma, 1996, p. 330.
29 Neumayer, 2013, p. 134.
30 Varma, 1984, p. 207; 2012, p. 62.
31 Tewari, 1990, pp. 12–13.
32 Neumayer, 1993, p. 115.
33 Tewari, 1987, p. 26.
34 Neumayer, 1983, p. 17.
35 Mathpal, 1984, pp. 87, 105.
36 Anand, 1991–92, p. 60.
37 Neumayer, 1993, p. 56.
38 Neumayer, 2013, p. 134.
39 Sonawane, 2002, p. 71.
40 Neumayer, 1983, p. 35.
41 Nath and Biswas, 1980, p. 122.
42 *Indian Archaeology 1967–68–A Review*, p. 28.
43 Thomas and Joglekar, 1994, p. 189.
44 Ibid.

CHAPTER 3 (pp. 44–59)

1 Meadow, 1989, p. 68; Jarrige, 1987–88, p. 187.
2 Casal, 1964, pp. 164–69.
3 Prashad, 1936, pp. 30–31.
4 Banerjee et al., 2003, pp. 267–339.
5 V.N. Prabhakar, personal communication.
6 Sharma, 1990, p. 381.
7 Thomas et al., 1995a, p. 38.
8 Thomas et al., 1996b, p. 304.
9 Nath and Rao, 1985, Vol. II, pp. 637, 642.
10 Thomas, 1977, p. 138.
11 Chattopadhyaya, 2002, pp. 393, 395.
12 Enzel et al., 2000, p. 235.
13 Manuel, 2004–05, pp. 21–26; 2008b, p. 172.
14 Lahiri, 2000, p. 12.
15 Rookmaaker, 2000, p. 66.
16 Meadow, 1993, p. 197.
17 Mackay, 1931a, Vol. I, p. 348.
18 Dales and Kenoyer, 1989, p. 20.
19 Mackay, 1931b, Vol. II, p. 387.
20 Mackay, 1938, Vol. I, p. 331.
21 Marshall, 1931, Vol. I, p. 70.
22 Vats, 1940, Vol. I, pp. 307–08.
23 Mackay, 1931b, Vol. II, pp. 387, 399.
24 Ibid., p. 398.
25 Marshall, 1931, Vol. I, pp. 54–55.
26 Sullivan, 1964, p. 122.
27 Atre, 1985–86, p. 9.
28 Marshall, 1931, Vol. I, p. 55.
29 Atre, 1990, p. 44.
30 Dhavalikar, 1993, p. 423.
31 Frankfort, 1965, p. 305.
32 Francfort, 1984, p. 303.
33 Chakravarty et al., 1989, p. 75.
34 Manuel, 2004–05, p. 23.
35 Kumar, 2000–01, pp. 25, 27.

36 Dhavalikar, 1984, p. 251.
37 Manuel, 2008a, p. 36.
38 Ibid.
39 Rookmaaker, 2002, p. 928.
40 Divyabhanusinh, personal communication.
41 Bautze, 1995, p. 28.
42 Dutta, 1991, pp. 23–25.
43 Bautze, 1985, pp. 416, 428.
44 Ibid., p. 428.
45 Sharma, 1969, pp. 70, 73.
46 Gupta, 1980, p. 53.
47 Chandra, 1970, p. 36.
48 Gupta, 1980, p. 58.
49 Marshall et al., 1982, Vol. I, p. 173.
50 Ibid.
51 Quintanilla, 2007, p. 76.
52 Ibid.
53 Bautze, 1995, p. 39.
54 Das, 2008, p. 93.
55 Nagar, 1949, p. 7.
56 Manuel, 2008a, p. 34.
57 *Indian Archaeology 1999–2000—A Review*, p. 155.
58 Bautze, 1985, p. 415.
59 Jamison, 1998, p. 252.
60 Griffith, 1957, p. 258.
61 Ibid., p. 264.
62 Jamison, 1998, p. 255.
63 *Sankhayana Shrautasutra* 14.33.26 (Bautze, 1985, p. 410).
64 *Jaiminiya Brahmana* II.103 (Jamison, 1998, p. 255).
65 Hare, 1947, pp. 6–11.
66 Francis, 1969, Vol. V, pp. 215–16.
67 Cowell and Rouse, 1969, Vol. VI, pp. 135–36.
68 Winternitz, 1977, Vol. II, pp. 174–75.
69 Davids, 1890, Part I, p. 38.
70 Jacobi, 1884, Part I, p. 261.
71 *Apastamba Dharmasutra* 1.17.37 (Olivelle, 2003, p. 55).
72 *Gautama Dharmasutra* 17.27 (Olivelle, 2003, p. 165).
73 *Baudhayana Dharmasutra* 1.12.5 (Olivelle, 2003, p. 223).
74 *Manava Dharmashastra* 5.17–18 (Olivelle, 2006, p. 139).
75 *Apastamba Dharmasutra* 2.17.1 (Olivelle, 2003, p. 99).
76 *Manava Dharmashastra* 3.272 (Olivelle, 2006, p. 122).
77 *Charaka Samhita* I.27.39 (Sharma, 1994, Vol. I, p. 197).
78 *Charaka Samhita* I.27.84 (Sharma, 1994, Vol. I, p. 201).
79 Olivelle, 2002, p. 25.
80 *Charaka Samhita* VI.8.154 (Sharma, 1994, Vol. II, p. 156).
81 *Charaka Samhita* IV.8.62 (Sharma, 1994, Vol. I, p. 486).
82 *Sushruta Samhita* I.46.49–50 (Bhishagratna, 1963, Vol. I, pp. [illegible]).
83 *Sushruta Samhita* I.46.51 (Bhishagratna, 1963, Vol. I, p. 488).
84 *Sushruta Samhita* I.46.53 (Bhishagratna, 1963, Vol. I, p. 489).
85 *Sundarakanda* 5.9.13–15ab (Goldman and Goldman, 1996, pp. 139–40).
86 Brockington, 1984, p. 93.
87 Griffith, 1963, Canto XI, p. 402; Shastri, 1957, Vol. II, p. 362.
88 Goldman and Goldman, 1996, p. 377.
89 *Adiparvan* 1.89.39 (Buitenen, 1973, Vol. I, p. 212).
90 Buitenen, 1973, Vol. I, p. 455.
91 *Karnaparvan* 8.11.6 (Bowles, 2006, Vol. I, p. 157).
92 *Karnaparvan* 8.6.37 (Bowles, 2006, Vol. I, p. 125).
93 *Anushasanaparvan* 13.88.10 (Ganguli, 1970, Vol. XI, p. 145).
94 *Arthashastra* 2.17.13 (Kangle, 1963, Part II, p. 149).
95 *Arthashastra* 2.18.12 (Kangle, 1963, Part II, p. 152).
96 *Arthashastra* 2.18.16 (Kangle, 1963, Part II, p. 152).
97 Mookerji, 1960, pp. 61–62.
98 Hultzsch, 1969, p. 127.
99 McCrindle, 1882, pp. 26–27.
100 McCrindle, 1896, p. 187.
101 Ibid., p. 216.
102 McCrindle, 1901, p. 59.
103 McCrindle, 1877, pp. 59–60.
104 Casson, 1989, p. 27.
105 McCrindle, 1901, p. 136.
106 Ibid., p. 145.

CHAPTER 4 (pp. 62–79)

1 Thackston, 1996, p. 275.
2 For further details on these copies: Smart, 1977. For the remnants of another incomplete or lost copy: Smart, 1987, pp. 105–15.
3 Painter Dhanraj: Suleiman, 1970, pl. 46.
4 Randhawa, 1983, pl. 121c.
5 No. IM 264-1913, unpublished.
6 No. RVI 1800 mounted on a page of the *Farhang-i-Jahangiri*.
7 Folios 351v –352r: Suleiman, 1970, pl. 51–52.
8 Randhawa, 1983, col. pl. XVII.
9 Thackston, 1996, p. 335.
10 Thackston, 1996, p. 275; Beveridge, 1921, p. 378.
11 For an illustration of this phenomenon in a late 16th or early 17th-century Iranian manuscript of *'Aja'ib al-Makhluqat* in the Walters Art Museum: Ettinghausen, 1950, pl. 22.
12 Suleiman, 1970, pl. 56.
13 Tyulayev, 1960, pl. 36; Vorobeva, 2005, f. 379a, pl. 56.
14 MAK B I 8770/52: Seyller et al., 2002, no. 49.
15 MAK B I 8770/17: Seyller et al., 2002, no. R134.
16 Maharaja Sawai Man Singh II Museum, Jaipur, *Razmnama*, painting no. AG1703, composition by Daswant and colouring by Kanha: Das, 1998, p. 66, fig. 13.
17 Birla Academy of Art and Culture, Kolkata, *Razmnama* f. 46r: Das, 2005, pl. 12.
18 Sotheby's London, October 19, 1994, lot 112. Present whereabouts unknown.
19 Ettinghausen, 1950, pl. 30: Isfandiyar killing *karg* monsters, from the *Shahnama* dated 1544 in the Princeton University Library.
20 Robinson, 1976, no. 378: Solomon and

the Queen of Sheba, Shiraz style, 16th century.

21 Titley, 1977, no. 110(25): Isfandiyar killing a rhinoceros, from a *Shahnama* dated 1614; no. 113(1): Suleiman enthroned surrounded by birds and animals including a rhinoceros, from a Safavid manuscript of c. 1590–95.

22 MS no. 36: Leach, 1995, II, 1040-1, no. 11.31: Gustaps kills a rhino; no. 11.34: Isfandiyar on horseback fights a rhinoceros, whose mate already lies dead.

23 No. Or 12208 f. 298a by Madhav Khanazad: Brend, 1995, fig. 39.

24 Majnu in the wilderness, attributed to Miskin, no. 1981.81: Cummins, 2006, pl. 17.

25 Majnu in the wilderness, signed Ma'sud or Maqsud 1612, No. 58.20/29: Goswamy, 1986, no. 100.

26 Majnu in the wilderness, c. 1630, Museum Rietberg, Alice Boner Bequest: Goswamy and Fischer, 1992, no. 89.

27 Topsfield, 2008, pl. 30 and 73.

28 Titley, 1977, no. 234–44, 246.

29 Freer Gallery of Art, no. 07.625: Ettinghausen, 1950, pl. 14b.

30 For a recent study on Salim's days at Allahabad: Das, 2009.

31 The painting was formerly in the Otto Sohn-Rethel collection, Dusseldorf: Ettinghausen, 1950, pl. 33; Pal, 1983, no. M45; Divyabhanusinh, 1999, pp. 106–07, fig. 9.

32 No. In 50.1: Leach, 1995, Vol. 32 I, 2.169; Wright et al., 2008.

33 Los Angeles County Museum of Art, no. M83.137: Pal, 1993, no. 65.

34 Art Gallery of New South Wales, Sydney: Divine and Courtly Life in Indian Painting, 4 October 1991–8 January 1992, Sydney 1991, no. 51, cover.

35 For an overview: Okada, 1992; Divyabhanusinh, 1999; Das, 2009, pp. 62–63.

36 Bautze, 1985, p. 422 note 43, no. 3.

37 Sotheby's London, The Sven Gahlin Collection sale, October 6, 2015, lot 16. I am thankful to Dan Ehnbom and Simon Ray for drawing my attention to this sale and for the supply of a high resolution image.

38 The attribution to Narsingh was made by Milo Beach. Details of Narsingh's work are given in the catalogue entry on p. 40 of the Sotheby's 2015 sale catalogue.

39 Thackston, 1999, p. 403.

40 Abu'l Fazl Allami, *A'in-i-Akbari*, II, p. 285.

41 Staatliche Museen zu Berlin, Preussischer Kulturbesitz, Museum fur Islamische Kunst, no. I 1986.127: Kroger, 2004, p. 247, fig. 9.

42 Private Collection: Jean Pozzi Sale, December 5, 1970, no. 77.

43 Foster, 1921, p. 246.

44 By unicorns Coryat meant rhinoceroses as per prevailing practice: Foster 1921, p. 246.

45 No. AG 969. For a note on this album: Das, 1995, pp. 71–85; Das, 2012, p. 88, fig. V.10 and p. 161 note 28. For the cassowary and other birds and animals in this album: ibid., pp. 152 and 165 note 16.

46 No. J4595 fol.4a: Kuhnel, 1933, pl. 22.

47 Soustiel and David, 1973, no. 59.

48 Most collections of Indian arms, armour and weapons in the country and abroad contain shields made of rhino hides, prized for their toughness and light weight. A superb highly decorated example in Maharaja Sawai Man Singh II Museum, Jaipur has recently been published: Elgood, 2015, pp. 162–63.

49 For a rare archer's ring made from walrus ivory with European religious figures carved on it for Jahangir probably during the years of his revolt at Allahabad and in the Maharaja Sawai Man Singh II Museum, Jaipur: Das, 1983; Elgood, 2015, pp. 230–34 (Dr Robert Elgood has made a detailed analysis of all available evidence and suggested very convincingly a date "probably 1619").

CHAPTER 5 (pp. 80–95)

1 Sarabhai, 1959, col. pl. Y; Khandalavala and Doshi, 1975, Vol. III, pl. 268A.

2 Khandalavala and Chandra, 1969, col. pl. 19b.

3 British Library Or 3299 folio 238a, *Karkadan*: Titley, 1977, no. 366.128.

4 Khandalavala and Chandra, 1974, fig. 42.

5 Present whereabouts of the folio not known; it was formerly in the Temple Collection, New York and later in the Ehrenfeld Collection, San Francisco: Hutchins, 1980, no. 15; Ehnbom, 1985, no. 2.

6 Beach, 1974, pl. LXX, fig. 74; for a coloured version: Welch and Beach, 1965, p. 72, no. 27; the drawing was first published in Beach, 1974, fig. 71.

7 Bautze, 1985.

8 Thackston, 1996, p. 403.

9 Khandalavala and Chandra, 1965, col. pl. H.

10 Rijksmuseum Amsterdam no. RP-T-1993-119: Catherine Benkaim, "The 'Stipple Master'", in Beach, Fischer and Goswamy, 2011, Vol. II, p. 522.

11 Baroda Museum and Painting Gallery no. PG5A.4203: Devkar, 1957, pl. XXII fig.7.

12 No. 540: Kala, 1961, pl. 8.

13 Acc. no. 64.1/4 reproduced in Gorakshkar, 1977, no. 379.

14 Topsfield, 2002, fig. 110.

15 V. Ball, tr., William Crooke ed., *Travels in India by Jean-Baptiste Tavernier, Baron of Aubonne,*

London, 1889, Vol. I, pp. 114–15, as quoted in Bautze, 1985, p. 426.

16 Bautze, 1985, fig. 5.

17 Heber, 1828, Vol. II, pp. 58–59; Vol. III, p. 5.

18 The atlas is in the British Library, India Office Library Collection APAC Or 4039: Gole, 1988. See also Rookmaaker, 2014, pp. 95–96, fig. 1–3.

19 Archer, 1992, no. 89 (13): V&A no. IS 25-1980.13; no. 89 (12): V&A no. IS 25-1980.12.

20 Yale Center for British Art no. B.2006.14.3: Shaffer, 2011, no. 140.

21 Shaffer, 2011, no. 141.

22 Parihar, 2015, pl. 2.39.

23 Bautze, 1985, fig. 7.

24 Michell, 1983, no. 401, p. 617; Haq, 2014, pl. 167, 213, 214.

25 Pal, 2013, pp. 116–17.

26 Carvalho, 2008.

27 *Embroidered Quilts from 24 the Museu Nacional de Arte Antiga: Exhibition Catalogue*, Kensington Palace, London, 1970, no. 4. A detail showing the rhinoceros.

28 National Gallery of Art, Widener Collection, no. 1942.9.475: Walker, 1997, cat. no. 10, fig. 48; Das, 2010.

29 Burrell Collection, Glasgow, inv. no. 1934/9/1: Cohen, 1996, p. 118, fig. 13. A fine example of a mythical tree with branches drawn in a similar manner with heads of birds, animals and clusters of animal heads and flowers, formerly in the Benkaim Collection, is now in the Cleveland Museum of Art no. 2013.319: first published in Welch and Zebrowski, 1973, no. 60.

30 Kauz, 2006, p. 45.

31 Abu'l Fazl Allami, *A'in-i-Akbari*, Vol. II, p. 285.

32 Foster, 1921, p. 176.

33 Ball (see note 15 above), Vol. II, pp. 319–20.

34 Bautze, 1985, p. 426, note 52. All efforts to get images of this rare object as well as the cup in the Indian Museum failed.

35 *Exotica: The Portuguese Discoveries and the Renaissance Kunstkammer*, Lisbon: Calouste Gulbenkian Foundation, 2001, no. 59.

CHAPTER 6 (pp. 98–113)

1 Bennett, 1984.

2 Williamson, 1807, Foreword.

3 Ibid., pp. 44–47.

4 Sterndale, 1886, pl. IX.

5 Inglis, 1888, pp. 298–99.

6 "Maori", 1878, p. 213.

7 Kinlock, 1885, p. 63.

8 Ibid.

9 Russell, 1900, p. 339.

10 Dollman and Burlace, 1935, p. 333.

11 Rookmaaker, 1998, p. 21.

12 Rookmaaker et al., 2016, p. 10.

13 Sharar, 1994, p. 54.

14 Ibid., p. 120.

15 Cooch Behar, 1908, pp. 1–2, 449.

16 Rookmaaker et al., 2016, p. 14.

17 Williamson, 1807, p. 45; "Maori", 1878, p. 218.

18 Baldwin, 1877, p. 145.

19 Kinlock, 1885, p. 61.

20 Aflalo, 1904, p. 61.

21 Eardley-Wilmot, 1910, p. 240.

22 Stebbing, 1920, p. 259; Stockley, 1928, p. 101; Alexander and Martin-Leake, 1932, p. 80.

23 Barthakur and Sahgal, 2005, p. 27.

24 Ibid., pp. 27–29; Gee, 1964, p. 154.

25 Barthakur and Sahgal, 2005, p. 31; Kakati, 2016, pp. 61–62.

26 National Archives of India, Home public, August 1904, No. 15.

27 *Times of India*, New Delhi, January 18, 2005.

28 Stracey, 1964, p. 152.

29 Saikia, 2005, p. 277.

30 Stracey, 1964, p. 98; Seshadri, 1986, p. 113.

31 Gee, 1964, p. 157.

32 Seshadri, 1969, p. 106.

33 Dharmakumarsinhji et al., 1970, pp. 35–36, 91–92.

34 Dutta, 1991, pp. 24–25.

35 Rookmaaker et al., 2016, p. 22.

CHAPTER 7 (pp. 116–129)

1 B.N. Basu, 1944, pp. 259–68.

2 Martin, 1983, p. 25.

3 Ibid., pp. 26–27.

4 *Hindustan Times*, New Delhi, November 1, 2017.

5 Dinerstein, 2003, p. 31

6 Gee, 1964, pp. 152–53.

7 Martin, 1983, p. 25; Ellis, 2005, p. 120–21.

8 Martin, 1983, pp. 47, 52, 69.

9 Ibid., p. 90.

10 Anonymous, 2015, p. 24.

11 Figures obtained from Kedar Gore of Corbett Foundation, personal communication, 2015; *Hindustan Times*, Mumbai, November 1, 2017; Traffic India, personal communication, 2018.

12 *Hindustan Times*, Mumbai, January 2, 2018.

13 Ibid.; Rookmaaker et al., 2016, p. 5.

14 Ranjitsinh, 1997, p. 98.

15 Dunn, 2015.

16 Misra with Ottaway, 2012, pp. 147, 184–89.

17 Ibid., pp. 186–91.

18 Ibid., pp. 145, 211–19.

19 Figures obtained by Dilpreet B. Chabbra of Traffic-India from government sources in India and Nepal; Rookmaaker et al., 2016, p. 12. The latter adds two more animals in Katarniaghat WLS (Uttar Pradesh) and three more in Parsa WR (Nepal).

20 Divyabhanusinh, 2008, No. 141–45, p. 78; Ranjitsinh, 1997, p. 98.

21 Singh and Rao, 1984, pp. 1–76.

22 Sharma and Gupta, 2015, pp. 2–3.

23 Ibid.

24 Figures obtained by Amit Sharma and Joydeep Bose of World Wide Fund for Nature–India from government sources.

25 Rookmaaker et al., 2016, p. 16.

Works marked * are additional references, not cited in the text.

A

Abu'l Fazl Allami. *A'in-i-Akbari*. English translation by H.S. Jarrett, corrected and annotated by J.N. Sarkar. Calcutta: Asiatic Society, 1949.

Aflalo, F.G. 1904. *The Sportsman's Book for India*. London: Horace Marshall & Son.

Alexander, R.D.T. and A. Martin-Leake. 1932. *Some Signposts to Shikar*. Calcutta: Fred R. Greyar for the Bengal Nagpur Railway Magazine.

Allaby, Michael. 1985. *The Oxford Dictionary of Natural History*. Oxford: Oxford University Press.

Anand, Kumar. 1991–92. "Rock Paintings of Kaimur", *The Journal of the Bihar Purāvid Pariṣad* XV and XVI: 55–64.

*Anonymous. 1934. "The Wild Animals of the Indian Empire, Part II", *Journal of the Bombay Natural History Society* XXXVII (1): 70–73.

Anonymous. 2015. *Conservation Programme Strategy*. New Delhi: World Wide Fund for Nature–India.

Apte, Vaman Shivram. 1992. *The Practical Sanskrit–English Dictionary*. Delhi: Motilal Banarsidass.

Archer, Mildred. 1992. *Company Paintings: Indian Paintings of the British Period*. London and Ahmedabad: Victoria and Albert Museum in association with Mapin Publishing.

Atre, Shubhangana. 1985–86. "Lady of Beasts", *Purātattva* 16: 7–14.

Atre, Shubhangana. 1990. "Harappan Seal Motifs and the Animal Retinue", *Bulletin of the Deccan College Postgraduate & Research Institute* 49: 43–49.

B

Badam, G.L. 1979. *Pleistocene Fauna of India with Special Reference to the Siwaliks*. Pune: Deccan College Postgraduate and Research Institute.

Badam, G.L. and S.C. Jayakaran. 1993. "Pleistocene Vertebrate Fossils from Tamil Nadu, India", in *Evolving Landscapes and Evolving Biotas of East Asia since the Mid-Tertiary*, ed. Nina G. Jablonski. Hong Kong: Centre of Asian Studies, The University of Hong Kong.

Baldwin, Capt. J.H. 1877. *The Large and Small Game of Bengal and North-Western Provinces of India*. London: Henry S. King & Co.

Banerjee, S., R.N. Mukherjee and B. Nath. 2003. "Identification of Animal Remains", in *Excavations at Kalibangan: The Early Harappans (1960–1969)*, ed. B.B. Lal, Jagat Pati Joshi, B.K. Thapar and Madhu Bala. *Memoirs of the Archaeological Survey of India No. 98*. New Delhi: Archaeological Survey of India, pp. 267–339.

Barthakur, Ranjit and Bittu Sahgal. 2005. *The Kaziranga Inheritance*. Mumbai: Sanctuary Asia.

Basu, B.N. 1944. *Kama Sutra of Vatsayana: The Hindu Art of Love*. Calcutta: Medical Books Company.

Basu, K.K., tr. 1944. *Yahiya Bin Ahmad Abdulla Sirhindi, The Tarikhi-i-Mubarakshahi*. Baroda: Oriental Institute.

Bautze, Joachim. 1985. "The Problem of the Khaḍga (*Rhinoceros unicornis*) in the Light of Archaeological Finds and Art", in *South Asian Archaeology 1983*, ed. J. Schotsman and M. Taddei. Naples: Istituto Universitario Orientale, pp. 405–33.

Bautze, Joachim. 1995. *Early Indian Terracottas*. Leiden, New York, Koln: E.J. Brill.

Beach, Milo Cleveland. 1974. *Rajput Paintings at Bundi and Kota*. Ascona: Artibus Asiae Publishers.

Beach, Milo C., Eberhard Fischer and B.N. Goswamy with Jorrit Britschgi, eds. 2011. *Masters of Indian Painting*, 2 vols. Zurich: Artibus Asiae Publishers.

Bennett, Scott. 1984. "Shikar and the Raj", *South Asia: Journal of South Asian Studies* New Series III (2): 72–88.

Beveridge, Annette S., tr. 1921. *The Babur-nama in English*, 2 vols. London: Luzac & Co.

Bhishagratna, Kaviraj Kunjalal. 1963. *An English Translation of the Sushruta Samhita Based on Original Sanskrit Text. Vol. I Sutra-Sthāna*. Varanasi: The Chowkhamba Sanskrit Series Office.

Blumenschine, R.J. and U.C. Chattopadhyaya. 1983. "A Preliminary Report on the Terminal Pleistocene Fauna of the Middle Son Valley", in *Palaeoenvironments and Prehistory in the Middle Son Valley*, ed. G.R. Sharma and J.D. Clark. Allahabad: Abinash Prakashan, pp. 281–84.

Bowles, Adam. 2006. *Mahabharata*, Book Eight: *Karṇa*, Vol. I. New York: Clay Sanskrit Library, New York University Press, JJC Foundation.

Brend, Barbara. 1995. *The Emperor Akbar's*

Khamsa of Nizami, London: The British Library.

Brockington, J.L. 1984. *Righteous Rama: The Evolution of an Epic*. Delhi: Oxford University Press.

Buitenen, J.A.B. van, tr. and ed. 1973. *The Mahābhārata* Book 1: *The Book of the Beginning*, Vol. I. Chicago and London: The University of Chicago Press.

C

Carvalho, Pedro Maura. 2008. *Luxury for Export: Artistic Exchanges between India and Portugal around 1600*. Boston: Isabella Stewart Gardner Museum.

Casal, Jean-Marie. 1964. *Fouilles D'Amri*. Vol. I: Texte. Paris: Librairie C. Klincksieck.

Casson, Lionel. 1989. *The Periplus Maris Erythraei: Text with Introduction, Translation, and Commentary*. Princeton, NJ: Princeton University Press.

Chakravarty, Kalyan Kumar, Vishnu Shridhar Wakankar and Maheshwari Dayal Khare. 1989. *Dangawada Excavations*. Bhopal: Commissioner, Archaeology and Museums, Madhya Pradesh.

Chandra, Pramod. 1970. *Stone Sculpture in the Allahabad Museum: A Descriptive Catalogue*. Poona: American Institute of Indian Studies.

Chattopadhyaya, U.C. 1996. "Settlement Pattern and the Spatial Organization of Subsistence and Mortuary Practices in the Mesolithic Ganges Valley, North-central India", *World Archaeology* 27(3): 461–76.

Chattopadhyaya, U.C. 1999. "The Ganges Neolithic", in *Encyclopaedia of Prehistory*, ed. M. Ember and P.A. Peregrine. New Haven: Human Relations Area Files Inc., pp. 127–32.

Chattopadhyaya, U.C. 2002. "Researches in Archaeozoology of the Holocene Period (Including the Harappan Tradition in India and Pakistan)", in *Archaeology and Interactive Disciplines: Indian Archaeology in Retrospect*, ed. S. Settar and R. Korisettar. New Delhi: Manohar, Vol. III, pp. 365–422.

Clarke, T.H. 1986. *The Rhinoceros from Dürer to Stubbs 1515–1799*. London/New York: Sotheby's/Harper & Row.

Clutton-Brock, Juliet. 1965. *Excavations at Langhnaj, Part II: The Fauna*. Poona: Deccan College Postgraduate and Research Institute.

Cockburn, John. 1883. "On the Recent Existence of *Rhinoceros indicus* in the North-Western Provinces and a Description of a Tracing of an Archaic Rock Painting from Mirzapore Representing the Hunting of this Animal", *Journal of the Asiatic Society of Bengal* 52(1): 56–64.

Cohen, Steven. 1996. "A Fearful Symmetry", *Hali Annual*: 104–35, 190–94.

Cooch Behar, The Maharaja of. 1908. *Thirty-Seven Years of Big Game Shooting in Cooch Behar, the Duars, and Assam: A Rough Diary*. Bombay: Times Press.

Cowell, E.B. and W.H.D. Rouse, tr. 1969. *The Jātaka or Stories of the Buddha's Former Births*, ed. E.B. Cowell. Vol. VI. London: Luzac & Company, Ltd.

Cummins, Joan. 2006. *Indian Painting from Cave Temples to the Colonial Period*. Boston: MFA Publications.

Curzon, The Marquis of Kedelston, K.G. 1925. *British Government in India: The Story of the Viceroys and Government Houses*. London: Cassel & Co.

D

Dales, George F. and J. Mark Kenoyer. 1989. *Preliminary Report on the Fourth Season (January 15–March 31, 1989) of Research at Harappa, Pakistan*. Berkeley/Madison: University of California/University of Wisconsin.

Das, Asok Kumar. 1983. "A Narwhal Ivory Archer's Ring", *Marg* 36(4): 81–83.

Das, Asok Kumar. 1995. "Notes on an Aurangzeb Period Album of Birds and Animals", in *Indian Painting: Essays in Honour of Karl J. Khandalavala*, ed. B.N. Goswamy and Usha Bhatia. New Delhi: Lalit Kala Akademi, pp. 71–85.

Das, Asok Kumar. 1998. "Daswant", in *Mughal Masters: Further Studies*, ed. Asok Kumar Das. Mumbai: Marg Publications.

Das, Asok Kumar. 2005. *Paintings of the Razmnama: The Book of War*. Ahmedabad: Mapin Publishing.

Das, Asok Kumar. 2009. "Salim's Taswirkhana", in *Allahabad: Where the Rivers Meet*, ed. Neelam Saran Gaur. Mumbai: Marg Publications, pp. 56–71.

Das, Asok Kumar. 2012. *Wonders of Nature: Ustad Mansur at the Mughal Court*. Mumbai: Marg Publications.

Das, Chhanda. 2008. "Further Discovery of a Plaque Representing *Rhinoceros unicornis* from Dum Dum Excavation, West Bengal: An Observation", in *Recent Researches in Indian Art and Iconography*, ed. B. Sahai. Delhi: Kaveri Books, pp. 93–95.

Das, Syamali. 2010. "The Mughal Scenic Animal Carpet in the National Gallery of Art, Washington DC", *Marg* 61(4): 56–61.

*Das Gupta, P.C. 1959. "Early Terracottas from Chandraketugarh", *Lalit Kala* 6: 45–52.

Dave, K.N. and Lokesh Chandra. 1953. *Indian Scientific Nomenclature of Mammals of India,*

Burma and Ceylon. Nagpur: International Academy of Indian Culture.

Davids, Rhys T.W., tr. 1890. *The Questions of King Milinda* Part I. *Sacred Books of the East*, ed. F. Max Müller, Vol. XXXV. Oxford: Clarendon Press.

Devkar, V.L. 1957. "Some Recently Acquired Miniatures in the Baroda Museum", *Bulletin of the Baroda Museum and Painting Gallery* XII.

Dharmakumarsinhji, R.S., Y.R. Ghorpade, Zafar Futehally, M. Krishnan, Hari Dang, V.B. Singh and K.S. Sankhala. 1970. *Wildlife Conservation in India: Report of the Expert Committee*. New Delhi: Indian Board for Wildlife.

Dhavalikar, M.K. 1984. "Sub Indus Cultures of Central and Western India", in *Frontiers of the Indus Civilization*, ed. B.B. Lal and S.P. Gupta. New Delhi: Books and Books.

Dhavalikar, M.K. 1993. "Daimabad Bronzes", in *Harappan Civilization: A Recent Perspective*, ed. G.L. Possehl. New Delhi: Oxford and IBH, pp. 422–26.

*Diderot, M. and M. D'Alembert. 1751–80. *Encyclopedie ou dictionaire raisonné des sciences des arts et des meteirs par une société de gens de letters*. Paris: Briasson. Copy accessed from the Special Collections: Rare Books, University of Louisville.

Dinerstein, Eric. 2003. *The Return of the Unicorns: The Natural History and Conservation of the Greater Indian One-Horned Rhinoceros*. New York: Columbia University Press.

Dinerstein, Eric. 2011. "Family Rhinocerotidae (Rhinoceroses)", in Wilson and Mittermier, pp. 144–81.

Dinerstein, Eric. 2015. "Greater One-horned Rhinoceros", in Johnsingh and Manjrekar, pp. 95–111.

Divyabhanusinh. 1999. "Hunting in Mughal Painting", in *Flora and Fauna in Mughal Art*, ed. Som Prakash Verma. Mumbai: Marg Publications, pp. 94–108.

Divyabhanusinh. 2006. *The End of a Trail: The Cheetah in India*. New Delhi: Oxford University Press.

Divyabhanusinh. 2008. *The Story of Asia's Lions*. Mumbai: Marg Publications.

Dollman, Gary and J.B. Burlace. 1935. *Rowland Ward's Records of Big Games*. London: Rowland Ward.

*Doniger, Wendy. 2009. *The Hindus: An Alternative History*. New York: Penguin Group (USA).

Dunn, James. 2015. *Mailonline*. http://www.dailymail.co.uk/news/article-3291773/ King George V-hunting Nepal–pictures–expedition-killed 60 rhinos–tigers, 1911.html.

Dutta, Arup Kumar. 1991. *Unicornis: The Great Indian One-horned Rhinoceros*. Delhi: Konark Publishers.

E

Eardley-Wilmot, Sainthill. 1910. *Forest Life and Sport in India*. London: Edward Arnold.

Ehnbom, Daniel J. 1985. *Indian Miniatures: The Ehrenfeld Collection*. New York: Hudson Hills Press in association with the American Federation of Arts.

Elgood, Robert. 2015. *Arms and Armour at the Jaipur Court: The Royal Collection*. New Delhi: Niyogi Books.

Ellis, Richard. 2005. *Tiger Bone and Rhino Horn*. Washington DC: Island Press/Shearwater Books.

*Ellison, Bernard C. 1925. *H.R.H. The Prince of Wales's Sport in India*. London: William Heinemann.

Enzel, Y., L.L. Ely, S. Mishra, R. Ramesh, R. Amit, B. Lazar, S.N. Rajaguru, V.R. Baker and A. Sandle. 2000. "High Resolution Holocene Environmental Changes in the Thar Desert, Northwestern India", in *The Decline and Fall of the Indus Civilization*, ed. Nayanjot Lahiri. New Delhi: Permanent Black, pp. 226–38.

Ettinghausen, Richard. 1950. *Studies in Muslim Iconography I: The Unicorn*. Washington DC: Freer Gallery of Art.

F

Finn, Frank. 1929. *Sterndale's Mammalia of India*. Calcutta: Thacker, Spink & Co.

Foster, William, ed. 1921. *Early Travels in India 1583–1619*. London: Oxford University Press.

Francfort, H.P. 1984. "The Harappan Settlement of Shortughai", in *Frontiers of the Indus Civilization*, ed. B.B. Lal and S.P. Gupta. New Delhi: Books and Books, pp. 301–10.

Francis, H.T., tr. 1969. *The Jātaka or Stories of the Buddha's Former Births*, ed. E.B. Cowell. Vol. V. London: Luzac & Company, Ltd.

Frankfort, Henri. 1965. *Cylinder Seals: A Documentary Essay on the Art and Religion of the Ancient Near East*. London: The Gregg Press.

G

Ganguli, Kisari Mohan, tr. 1970. *The Mahabharata of Krishna-Dwaipayana Vyasa*, Vol. XI: *Anusasana Parva*, Part II. New Delhi: Munshiram Manoharlal.

*Gee, E.P., 1953. "The Life History of the Great Indian One-horned Rhinoceros", *Journal of the Bombay Natural History Society*, 51(2): 341–48; "Further Observations", 51(4): 765–72.

Gee, E.P. 1964. *The Wildlife of India*. London: Collins.

Gibb, H.A.R. tr. 1971. *The Travels of Ibn Battuta* A.D. *1325–1354*, Vol. III. Cambridge: Hakluyt Society.

Goldman, Robert P. and Sally J. Sutherland Goldman, tr. 1996. *The Rāmāyaṇa of Vālmīki: An Epic of Ancient India, Vol. V: Sundarakāṇḍa.* Princeton, NJ: Princeton University Press.

Gole, Susan. 1988. *Maps of Mughal India: Drawn by Colonel Jean-Baptiste-Joseph Gentil Agent for the French Government to the Court of Shuja-ud-daula at Faizabad in 1770.* Delhi: Manohar.

Gorakshkar, S. 1977. *Animals in Indian Art.* Bombay: Prince of Wales Museum.

Goswamy, B.N. 1986. *Essence of Indian Art.* San Francisco: Asian Art Museum.

Goswamy, B.N. and Eberhard Fischer. 1992. *Wonders of a Golden Age.* Zurich: Museum Rietberg.

Gotch, A.F. 1995. *Latin Names Explained: A Guide to the Scientific Classification of Reptiles, Birds & Animals.* London: Blanford/Cassell.

Griffith, Ralph T.H., tr. 1957. *The Texts of the White Yajurveda.* Banaras: E.J. Lazarus & Co.

Griffith, Ralph T.H., tr. 1963. *The Ramayan of Valmiki.* Varanasi: The Chowkhamba Sanskrit Series Office.

Gupta, S.P. 1980. *The Roots of Indian Art.* Delhi: B.R. Publishing Corporation.

H

Haq, Zulekha. 2014. *Terracottas of Bengal: An Analytical Study.* Dhaka: International Centre for the Study of Bengal Art.

Hare, E.M., tr. 1947. *Woven Cadences of Early Buddhists (Sutta-Nipāta).* London: Oxford University Press.

Heber, Reginald. 1828. *Narrative of a Journey through the Upper Provinces through Calcutta to Bombay...1824–25*, 3 vols., London. Reprint, New Delhi and Madras: Asian Educational Services.

Hultzsch, E. 1969. *Corpus Inscriptionum Indicarum: Inscriptions of Aśoka*, Vol. I. Delhi: Indological Book House.

Hutchins, Francis. 1980. *Young Krishna.* New Hampshire: Amanda Press.

I

Indian Archaeology—A Review. 1967–68; 1971–72; 1999–2000.

Inglis, The Hon. James. 1888. *Tent Life in Tiger Land: Being Sporting Reminiscences of a Pioneer Planter in an Indian Frontier District.* London: Sampson Low, Marton, Searle & Rivington.

Isenberg, Andrew C. 2001. *The Destruction of the Bison.* Cambridge: Cambridge University Press.

J

Jacobi, Hermann, tr. 1884. *Jaina Sūtras*, Part I. *Sacred Books of the East*, ed. F. Max Müller, Vol. XXII. Oxford: Clarendon Press.

Jamison, Stephanie W. 1998. "Rhinoceros Toes, Manu V.17–18, and the Development of the Dharma System", *Journal of the American Oriental Society* 118(2): 249–56.

Jarrige, J.F. 1987–88. "Excavations at Nausharo", *Pakistan Archaeology* 23: 149–203.

Jerdon, T.C. 1874. *The Mammals of India: A Natural History of all the Animals Known to Inhabit Continental India.* London: John Weldon.

Joglekar, P.P., V.D. Misra, J.N. Pal and M.C. Gupta. 2003. *Mesolithic Mahadaha: The Faunal Remains.* Allahabad: University of Allahabad.

Johnsingh, A.J.T. and Nima Manjrekar, eds. 2015. *The Mammals of South Asia*, Vol. II. Hyderabad: Universities Press.

*Joshi, Jagat Pati and Asko Parpola, eds. 1987. *Corpus of Indus Seals and Inscriptions: 1. Collections in India.* Helsinki: Suomalainen Tiedeakatemia.

K

Kakati, Kashmira, 2016. "Wildlife of the Brahmaputra", Final Report to FREMAA (Flood and River Erosion Management Agency of Assam), for ADB AIFRERMEP assisted Project in *The Mahabahu Brahmaputra*, ed. K.K. Dwivedi, Guwahati: Flood and River Erosion Management Agency of Assam.

Kala, S.C. 1961. *Indian Miniatures in the Allahabad Museum.* Allahabad: Allahabad Nagar Mahapalika.

Kangle, R.P. 1963. *The Kauṭilīya Arthaśāstra*, Part II. *An English Translation with Critical and Explanatory Notes.* Bombay: University of Bombay.

Kauz, Ralph. 2006. "Bengal Textiles as Tribute Items to Ming China", in *Textiles from India: The Global Trade*, ed. Rosemary Crill. Kolkata: Seagull Books.

Khandalavala, Karl and Moti Chandra. 1965. *Miniatures and Sculptures from the Collection of the Late Sir Cowasji Jehangir, Bart.* Bombay: Prince of Wales Museum.

Khandalavala, Karl and Moti Chandra. 1969. *New Documents of Indian Painting: A Reappraisal.* Bombay: Prince of Wales Museum.

Khandalavala, Karl and Moti Chandra. 1974. *An Illustrated Aranyaka Parvan in the Asiatic Society of Bombay.* Bombay: Prince of Wales Museum.

Khandalavala, Karl and Saryu Doshi. 1975. "Miniature Paintings", in *Jaina Art and*

Architecture, Vol. III, ed. A. Ghosh. Delhi: Bharatiya Jnanapitha

Kinlock, Col. Alexander A. 1885. *Large Game Shooting in Tibet, The Himalayas, and Northern India*. Calcutta: Thacker, Spink & Co.

Kroger, Jens. 2004. "On Maumud b. Ishaq al-Shihabi's Manuscript of *Yusuf va Zulaykha* of 964 (1557)", in *Muqarnas, Essays in Honor of J.M. Rogers*, ed. G. Necipoğlu, D. Beherens-Abouseif and Anna Contadini. Leiden: Brill.

Kuhnel, E. 1933. *Indische Miniaturen*. Berlin: Islamische Kunstabteilung.

Kumar, Giriraj. 2000–01. "Chronology of Indian Rock Art: A Fresh Attempt", *Purākalā* 11–12: 5–35.

L

Lahiri, Nayanjot. 2000. *The Decline and Fall of the Indus Civilization*. New Delhi: Permanent Black.

Laurie, William Andrew. 1978. *The Ecology and Behaviour of the Greater One-horned Rhinoceros*. PhD thesis. Cambridge: Selwyn College.

Laurie, W.A., E.M. Lang and C.P. Groves. 1983. "Rhinoceros unicornis", *Mammalian Species* 211: 1–6.

Leach, Linda York. 1995. *Mughal and Other Indian Paintings from the Chester Beatty Library*, 2 vols. London: Scorpion Cavendish.

Lukacs, John R., J.N. Pal and V.D. Misra. 1996. "Chronology and Diet in Mesolithic North India: A Preliminary Report of New AMS C Dates, δ C Isotope Values, and Their Significance", in *Bioarchaeology of Mesolithic India: An Integrated Approach, Colloquium XXXII of the International Union of Prehistoric and Protohistoric Sciences*, ed. G.E. Afanas'ev, S. Cleuziou, J.R. Lukacs and M. Tosi. Forli: ABACO Edizioni, pp. 301–11.

Lydekker, R. 1924. *The Game Animals of India, Burma, Malaya and Tibet*. London: Rowland Ward.

M

Mackay, Christine. 1997. "Works of Art on Paper: Conservation of the Study Collection and the Role of Facsimile", in *The Victoria Memorial Hall Calcutta: Conception, Collections, Conservation*, ed. Philippa Vaughan. Mumbai: Marg Publications, pp. 91–100.

Mackay, Ernest. 1931a. "Figurines and Model Animals", in Marshall, ed., 1931, Vol. I, pp. 338–55.

Mackay, Ernest. 1931b. "Seals, Seal Impressions, and Copper Tablets, with Tabulation", in Marshall, ed., 1931, Vol. II, pp. 370–405.

*Mackay, Ernest. 1931c. "Ivory, Shell, Faience, and Other Objects of Technical Interest", in Marshall, ed., 1931, Vol. II, pp. 562–88.

Mackay, Ernest. 1938. *Further Excavations at Mohenjo-daro: Being an Official Account of Archaeological Excavations at Mohenjo-daro Carried out by the Government of India between the Years 1927 and 1931*. 2 vols. Delhi: Manager of Publications.

*Mackay, Ernest. 1943. *Chanhu-Daro Excavations 1935–36*, ed. W.N. Brown. American Oriental Series. New Haven, CT: American Oriental Society.

Manuel, Joseph. 2004–05. "Harappan Environment as One Variable in the Preponderance of Rhinoceros and Paucity of Horse", *Purātattva* 35: 21–26.

Manuel, Joseph. 2008a. "Depiction of Rhinoceros: Transition from Popular Art to State Sponsored Art", in *Expressions in Indian Art: Essays in Memory of Shri M.C. Joshi*, ed. B.R. Mani and A. Tripathi. Delhi: Agam Kala Prakashan, Vol. I, pp. 33–38.

Manuel, Joseph. 2008b. "Why Deny Evidence on Horse?", in *Rock Art & Archaeology of India: Prof. Shankar Tiwari Commemoration Volume*, ed. K.K. Chakravarty and G.L. Badam. Delhi: Agam Kala Prakashan, pp. 169–77.

"Maori". 1878. *Sport and Work on the Nepaul Frontier or Twelve Years Sporting Reminiscences of an Indigo Planter*. London: Macmillan.

Marshall, Sir John, ed. 1931. *Mohenjo-Daro and the Indus Civilization: Being an Official Account of Archaeological Excavations at Mohenjo-Daro Carried out by the Government of India between the Years 1922 and 1927*. 3 vols. London: Arthur Probsthain.

Marshall, Sir John, A. Foucher and N.G. Majumdar. 1982. *The Monuments of Sanchi*, Vol. I. Delhi: Swati Publications.

Martin, Esmond Bradly. 1983. *Rhino Exploitation: The Trade in Rhino Products in India, Indonesia, Malaysia, Burma, Japan & South Korea*. Hongkong: World Wildlife Fund.

Mathpal, Yashodhar. 1984. *Prehistoric Rock Paintings of Bhimbetka*. New Delhi: Abhinav Publications.

McCrindle, J.W. 1877. *Ancient India as Described by Megasthenes and Arrian*. London: Trubner & Co.

McCrindle, J.W. 1882. *Ancient India as Described by Ktesias the Knidian*. London: Trubner & Co.

McCrindle, J.W. 1896. *Invasion of India by Alexander the Great*. Westminster: Archibald Constable and Co.

McCrindle, J.W. 1901. *Ancient India as Described in Classical Literature*. Westminster: Archibald Constable and Co.

Meadow, Richard H. 1989. "Continuity and Change in the Agriculture of the Greater Indus Valley: The Palaeoethnobotanical and Zooarchaeological Evidence", in *Old Problems and New Perspectives in the Archaeology of South Asia*, ed. J.M. Kenoyer. *Wisconsin Archaeological Reports 2*. Madison: University of Wisconsin, pp. 61–74.

Meadow, Richard H. 1993. "The Past, Present and Future of Bioarchaeological Studies in Pakistan with Specific Reference to Mohenjodaro and the Indus Civilization", *Pakistan Archaeology* 28: 183–215.

Michell, George, ed. 1983. *Brick Temples of Bengal: From the Archives of David McCutchion*. Princeton, NJ: Princeton University Press.

Misra, Hemanta with Jim Ottaway, Jr. 2012. *The Soul of the Rhino: A Nepali Adventure with Kings and Elephant Drivers, Bureaucrats and Scientists, and the Indian Rhinoceros*. New Delhi: Penguin Books.

Misra, V.D., J.N. Pal and M.C. Gupta. 2003. "Mesolithic Mahadaha: General Features", in *Mesolithic Mahadaha: The Faunal Remains*, by P.P. Joglekar, V.D. Misra, J.N. Pal and M.C. Gupta. Allahabad: University of Allahabad, pp. 1–37.

Momin, K.N., D.R. Shah, and G.M. Oza. 1973. "Great Indian Rhinoceros Inhabited Gujarat", *Current Science* 42(22): 801–03.

Monier-Williams, Sir Monier. 1985. *English–Sanskrit Dictionary*. New Delhi: Munshiram Manoharlal.

Monier-Williams, Sir Monier. 1988. *Sanskrit–English Dictionary*. New Delhi: Munshiram Manoharlal.

Mookerji, Radha Kumud. 1960. *Chandragupta Maurya and His Times*. Delhi: Motilal Banarsidass.

*Morris, P.A. 2006. *Van Ingen & Van Ingen: Artists in Taxidermy*. Ascot: MPM Publishing.

N

Nagar, M.M. 1949. "A Rhinoceros-Slayer Type Coin of Kumaragupta I", *Journal of the Numismatic Society of India* 11(1): 7–8.

Nair, P. Thankappan. 1977. *The Peacock: The National Bird of India*. Calcutta: Firma K.L.M.

Nath, Bhola and M.K. Biswas. 1980. "Animal Remains from Chirand, Saran District, Bihar", *Records of the Zoological Survey of India* 76: 115–24.

Nath, Bhola and G.V. Sreenivasa Rao. 1985. "Animal Remains from Lothal Excavations", in *Lothal: A Harappan Port Town 1955–62*, ed. S.R. Rao. Vol. II. *Memoirs of the Archaeological Survey of India No. 78*. New Delhi: Archaeological Survey of India, pp. 636–50.

Neumayer, Erwin. 1983. *Prehistoric Indian Rock Paintings*. Delhi: Oxford University Press.

Neumayer, Erwin. 1993. *Lines on Stone: The Prehistoric Rock Art of India*. New Delhi: Manohar.

Neumayer, Erwin. 2013. *Prehistoric Rock Art of India*. New Delhi: Oxford University Press.

O

OED. 1977. *The Compact Edition of the English Dictionary*. Oxford: Clarendon Press.

Okada, Amina. 1992. "Le prince Salīm à la chasse: Une miniature inédite peinte à Allāhābād", *Artibus Asiae*, LII: 3–4.

Olivelle, Patrick. 2002. *Food for Thought: Dietary Rules and Social Organization in Ancient India*. Amsterdam: Royal Netherlands Academy of Arts and Sciences, pp. 5–38.

Olivelle, Patrick. 2003. *Dharmasūtras: The Law Codes of Āpastamba, Gautama, Baudhāyana and Vasiṣṭḥa*. Delhi: Motilal Banarsidass.

Olivelle, Patrick. 2006. *Manu's Code of Law: A Critical Edition and Translation of the Mānava-Dharmaśāstra*. New Delhi: Oxford University Press.

P

Pal, J.N. 2002. "The Mesolithic Phase in the Ganga Valley", in *Recent Advances in Indian Archaeology*, ed. K. Padayya. New Delhi: Munshiram Manoharlal, pp. 69–80.

Pal, P. 1983. *Court Paintings of India*. New York: Ravi Kumar.

Pal, P. 1993. *Indian Painting: A Catalogue of the Los Angeles County Museum of Art Collection, Vol. 1, AD 1000–1700*. Los Angeles: Los Angeles County Museum of Art.

Pal, P. 2013. "The Mystery of Tulsiram's Durga and Ivory-Carving of Murshidabad", in *Murshidabad: Forgotten Capital of Bengal*, ed. Neeta Das and Rosie Llewellyn-Jones. Mumbai: The Marg Foundation, pp. 106–22.

Parihar, Subhash. 2015. *Islamic Architecture of Punjab (1206–1707)*. New Delhi: Aryan Books.

Pastoureau, Michel. 2011. *The Bear: History of a Fallen King*. Cambridge, MA: Bellcamp Press of Harvard University Press.

Prashad, B. 1936. *Animal Remains from Harappa. Memoirs of the Archaeological Survey of India No. 51*. Delhi: Manager of Publications.

Prater, S.H. 1948. *The Book of Indian Animals*. Bombay: Bombay Natural History Society.

Q

Quintanilla, Sonya Rhie. 2007. *History of Early Stone Sculpture at Mathura: ca. 150 BCE–100 CE*. Leiden and Boston: Brill.

R

Randhawa, M.S. 1983. *Paintings of the Babur Nama*. New Delhi: National Museum.

Ranjitsinh, M.K. 1997. *Beyond the Tiger: Portrait of Asia's Wildlife*. New Delhi: Brijbasi.

Ridley, Glynis. 2004. *Clara's Grand Tour: Travels with a Rhinoceros in Eighteenth Century Europe*. New York: Grove Press.

Robinson, Basil. 1976. *Persian Painting in the India Office Library*. London: Sotheby Parke-Bernet.

Rookmaaker, Kees. 1997. "Record of the Sundarbans Rhinoceros (*Rhinoceros sondaicus inermis*) in India and Bangladesh", *Pachyderm* 24 (July–December).

Rookmaaker, Kees. 2014. "Three Rhinos on Maps of India in Faizabad in the 18th Century", *Pachyderm* 55 (January).

Rookmaaker, K., A. Sharma, J. Bose, K. Thapa, D. Dutta, B. Jeffries, A.C. Williams, D. Ghose, M. Gupta, and S. Tornikoski. 2016. *The Greater One-Horned Rhino: Past, Present and Future*. Gland, Switzerland: WWF.

*Rookmaaker, L.C. 1978. "Two Collections of Rhinoceros Plates Compiled by James Douglas and James Parsons in the Eighteenth Century", *Journal of the Society for the Bibliography of Natural History* 9(1).

Rookmaaker, L.C. 1984. "The Former Distribution of the Indian Rhinoceros (*Rhinoceros unicornis*) in India and Pakistan", *Journal of the Bombay Natural History Society* 80(3): 555–63.

Rookmaaker, L.C. 1998. *The Rhinoceros in Captivity: A List of 2439 Rhinoceroses kept from Roman Times to 1994*. The Hague: SPB Academic Publishing Co.

Rookmaaker, L.C. 1999. "Records of the Rhinoceros in Northern India", *Saugetierkundliche Mitteilungen* 44(2): 51–78.

Rookmaaker, L.C. 2000. "Records of the Rhinoceros in Pakistan and Afghanistan", *Pakistan Journal of Zoology* 32(1), 65–74.

Rookmaaker, L.C. 2002. "Historical Records of the Rhinoceros (*Rhinoceros unicornis*) in Northern India and Pakistan", *Zoos Print Journal* 17(11): 923–29.

Rousselet, Louis. 1882. *India and Its Native Princes: Travels in Central India and in Presidencies of Bombay and Bengal*. London: Bickers & Sons.

Russell, C.E.M. 1900. *Bullet and Shot in Indian Forest, Plain and Hill*. London: W. Thacker and Co.

S

Sachau, Edward C., ed. 1983. *Alberuni's India: An Account of the Religion, Philosophy, Literature, Geography, Chronology, Astronomy, Customs, Laws and Astrology of India about AD 1030*. Vol. I. New Delhi: Munshiram Manoharlal.

Saikia, Arupjyoti. 2005. *Jungles, Reserves, Wildlife: A History of Forest Reserves in Assam*. Guwahati: Wildlife Area Development and Welfare Trust, Assam.

Sankalia, H.D. 1965. *Excavations at Langhnaj 1944–63, Part 1: Archaeology*. Poona: Deccan College Postgraduate and Research Institute.

Sanyal, Ram Brahma. 1892. *A Handbook of the Management of Animals in Captivity in Lower Bengal*. Calcutta: Bengal Secretariat Press.

Sarabhai, Manilal Nawab. 1959. *The Oldest Rajasthani Paintings from Jain Bhandaras*, Ahmedabad: Sarabhai Manilal Nawab.

Seshadri, Balakrishna. 1969. *The Twilight of India's Wildlife*. London: John Baker.

Seshadri, Balakrishna. 1986. *India's Wildlife and Wildlife Reserves*. New Delhi: Sterling Publishing.

Seyller, John, et al. 2002. *The Adventures of Hamza: Painting and Storytelling in Mughal India*. Washington DC: Freer Gallery of Art and Arthur M. Sackler Gallery.

Shaffer, Holly. 2011. *Adapting the Eye: An Archive of the British in India 1770–1830*. New Haven, CT: Yale Center for British Art.

Shah, D.R. 1980. "Animal Remains from Kanewal", in *Excavation at Kanewal*, by R.N Mehta, K.N. Momin and D.R. Shah. Vadodara: Department of Archaeology and Ancient History, M.S. University of Baroda, pp. 74–76.

*Shah, Sayid Ghulam Mustafa and Asko Parpola, eds. 1991. *Corpus of Indus Seals and Inscriptions: 2. Collections in Pakistan*. Helsinki: Suomalainen Tiedeakatemia.

Sharar, Abdul Halim. 1994. *Lucknow: The Last Phase of an Oriental Culture*, tr. and ed. E.S. Harcourt and Fakhir Hussain. Delhi: Oxford University Press.

Sharma, A.K. 1990. "Animal Bone Remains", in *Excavation at Surkotada 1971–72 and Exploration in Kutch*, ed. Jagat Pati Joshi. *Memoirs of the Archaeological Survey of India No. 87*. New Delhi: Archaeological Survey of India, pp. 372–83.

Sharma, G.R. 1969. *Excavations at Kausambi 1949–50. Memoirs of the Archaeological Survey of India No. 74*. Delhi: Manager of Publications.

Sharma, Priyavrat, ed. and tr. 1994. *Caraka-Saṃhitā: Agniveśa's Treatise Refined and Annotated by Caraka and Redacted by Dṛḍhabala* (text with English translation), Vol. I (*Sūtrasthāna to Indriyasthāna*) and Vol. II (*Chikitsāsthānam to Siddhisthānam*). Varanasi: Chaukhamba Orientalia.

Sharma, R. and M. Gupta. 2015. *Status and Monitoring of Greater One-horned Rhinoceros in Dudhwa National Park*. New Delhi: World Wide Fund for Nature–India.

Shastri, Hari Prasad. 1957. *The Ramayana of Valmiki*, Vol. II. London: Shanti Sadan.

Singh, Samar and Kishore Rao. 1984. *India's Rhino Re-introduction Programme*. New Delhi: Department of Environment, Government of India.

Smart, Ellen. 1977. *Paintings from the Baburnama: A Study of Sixteenth Century Mughal Manuscript Illustration*. PhD thesis. University of London.

Smart, Ellen. 1987. "Yet Another Illustrated Akbari *Baburnama* Manuscript", in *Facets of Indian Art: A Symposium Held at the Victoria and Albert Museum*, ed. R. Skelton et al. London: V&A Publications, pp. 105–15.

Smith, S.J., P. Mc Gehee, W.C. Mathews, eds. 1989. *Rowland Ward's Records of Big Game*. San Antonio, TX: Rowland Ward Publications/ Game Conservation International.

Sonawane, V.H. 2002. "Rock Art of Gujarat: A Regional Study", *Purākalā* 13(1–2): 67–85.

Soustiel, Jean and Marie-Christine David. 1973. *Miniatures orientales de l'Inde*. Paris: Editions Jean Soustiel.

Stebbing, E.P. 1920. *The Diary of a Sportsman Naturalist in India with Illustrations, Photographs and Sketches by the Author and Others*. London: John Lane, The Bodley Head.

Sterndale, R.A. 1884. *Natural History of India and Ceylon*. Calcutta: Thacker, Spink & Co.

Sterndale, R.A. 1886. *Denizens of the Jungles: A Series of Sketches of Wild Animals Illustrating their Forms and Natural Attitude*. Calcutta: Thacker, Spink & Co.

Stockley, Lt. Col. C.H. 1928. *Big Game Shooting in the Indian Empire*. London: Constable & Co.

Stracey, P.D. IFS. 1964. *Wild Life in India: Its Conservation and Control*. New Delhi: Ministry of Food and Agriculture.

Sukumar, Raman. 2011. *The Story of Asia's Elephants*. Mumbai: The Marg Foundation.

Suleiman, Hamid. 1970. *Miniatures of Babur-Nama*. Samarqand: Fan Publishing House of the Uzbek SSR.

Sullivan, H.P. 1964. "A Re-examination of the Religion of the Indus Civilization", *History of Religion* 4(1): 115–25.

T

Tewari, Rakesh. 1987. "Rhino-hunt Scene of the Ghora Mangara Rock-shelter, Mirzapur: A Reappraisal", *Bulletin of Museums & Archaeology in U.P.* 39: 25–29.

Tewari, Rakesh. 1990. *Rock Paintings of Mirzapur*. Lucknow: U.P. State Archaeological Organization.

Thackston, Wheeler M., ed. and tr. 1996. *The Baburnama: Memoirs of Babur, Prince and Emperor*. New York: Oxford University Press.

Thackston, Wheeler M., ed. and tr. 1999. *The Jahangirnama: Memoirs of Jahangir, Emperor of India*. Washington DC/New York: Freer and Sackler Gallery/Oxford University Press.

Thomas, P.K. 1977. *Archaeozoological Aspects of the Prehistoric Cultures of Western India*. PhD thesis. Poona: Deccan College.

Thomas, P.K. and P.P. Joglekar. 1994. "Holocene Faunal Studies in India", *Man and Environment* XIX(1–2): 179–203.

Thomas, P.K., P.P. Joglekar, Arati Deshpande-Mukherjee and S.J. Pawankar. 1995a. "Harappan Subsistence Patterns with Special Reference to Shikarpur, a Harappan Site in Gujarat", *Man and Environment* XX(2): 33–41.

Thomas, P.K., P.P. Joglekar, V.D. Mishra, J.N. Pandey and J.N. Pal. 1995b. "A Preliminary Report of the Faunal Remains from Damdama", *Man and Environment* XX(1): 29–36.

*Thomas, P.K., P.P. Joglekar, V.D. Mishra, J.N. Pandey and J.N. Pal. 1996a. "Faunal Evidence for the Mesolithic Food Economy of the Gangetic Plain with Special Reference to Damdama", in *Bioarchaeology of Mesolithic India: An Integrated Approach, Colloquium XXXII of the International Union of Prehistoric and Protohistoric Sciences*, ed. G.E. Afanas'ev, S. Cleuziou, J.R. Lukacs and M. Tosi. Forli: ABACO Edizioni, pp. 255–66.

Thomas, P.K., Yoshiyuki Matsushima and Arati Deshpande-Mukherjee. 1996b. "Faunal Remains", in *Kuntasi: A Harappan Emporium on West Coast*, ed. M.K. Dhavalikar, M.R. Raval and Y.M. Chitalwala. Pune: Deccan College Postgraduate and Research Institute, pp. 297–307.

Thompson, D'Arcy Wentworth. 1956. *The Works of Aristotle, Vol. IV: Historia Animalium*. Oxford: Oxford University Press.

Titley, Norah M. 1977. *Miniatures from Persian Manuscripts: A Catalogue and Subject Index of Paintings from Persia, India and Turkey in the British Library and the British Museum*. London: The British Library.

*Topsel, Edward. 1658. *The History of Four-Footed Beasts and Serpents...Collected out of the Writings of Conradus Gesner and Other Authors*. London. Copy accessed from the Special Collections: Rare Books, University of Louisville.

Topsfield, Andrew. 2002. *Court Painting at Udaipur: Art under the Patronage of the*

Maharanas of Mewar. Zurich: Artibus Asiae.

Topsfield, Andrew. 2008. *Paintings from Mughal India*. Oxford: Bodleian Library.

Tyulayev, S. 1960. *Miniatures of Babur Namah*. Moscow: State Fine Arts Publishing.

V

*Vaidya, Suresh. 1958. *Ahead Lies the Jungle*. London: Robert Hale.

Varma, Radha Kant. 1984. "The Rock-Art of Southern Uttar Pradesh with Special Reference to Mirzapur", in *Rock-Art of India: Paintings and Engraving*, ed. K.K. Chakravarty. New Delhi: Arnold-Heinemann, pp. 206–13.

Varma, Radha Kant. 1996. "Subsistence Economy of the Mesolithic Folk as Reflected in the Rock-paintings of the Vindhyan Region", in *Bioarchaeology of Mesolithic India: An Integrated Approach, Colloquium XXXII of the International Union of Prehistoric and Protohistoric Sciences*, ed. G.E. Afanas'ev, S. Cleuziou, J.R. Lukacs and M. Tosi. Forli: ABACO Edizioni, pp. 329–39.

*Varma, Radha Kant. 2012. *Rock Art of Central India: North Vindhyan Region with Special Reference to Mirzapur and the Adjoining Regions in Uttar Pradesh and Baghelkhand in Madhya Pradesh*. New Delhi: Aryan Books.

Vats, Madho Sarup. 1940. *Excavations at Harappa: Being an Account of Archaeological Excavations at Harappa Carried out between the Years 1920–21 and 1933–34*. 2 vols. Delhi: Manager of Publications.

Vogel, Jean Phillipe. 1962. *The Goose in Indian Art and Literature*. Leiden: E.J. Brill.

Vorobeva, S. 2005. *Baburnameh Miniatures from the Collection of the State Museum of Oriental Art*. Samara: Izadataelskii dom ("Agni").

W

Wakankar, V.S. 1973. *Painted Rock Shelters of India*. PhD thesis. Poona: Deccan College.

Walker, Brett L. 2005. *The Last Wolves of Japan*. Seattle: University of Washington Press.

Walker, Daniel. 1997. *Flowers Underfoot: Indian Carpets of the Mughal Era*. New York: Metropolitan Museum of Art.

Welch, Stuart Cary and Milo C. Beach. 1965. *Gods, Thrones and Peacocks*. New York: The Asia Society.

Welch, Stuart Cary and Mark Zebrowski. 1973. *A Flower from Every Meadow*. New York: The Asia Society.

Williamson, Capt. Thomas. 1807. *Oriental Field Sports: Being a Complete, Detailed and Accurate Description of the Wild Sports of the East*. London: Edward Orm.

*Wilcox, James. 1904. *From Kabul to Kumassi*. London: John Murray.

Wilson, Don E. and Russell A. Mittermeier. 2011. *Handbook of the Mammals of the World, Vol. 2: Hoofed Mammals*. Barcelona: Lynx Edicions.

Winternitz, Maurice. 1977. *A History of Indian Literature*, Vol. II. New Delhi: Oriental Books Reprint Corporation.

*Woodyatt, Maj. Gen. Nigel. 1923. *My Sporting Memories: Forty Years with Note Book and Gun*. London: Robert Jenkins.

Wright, Elaine et al. 2008. *Muraqqa: Imperial Mughal Albums from the Chester Beatty Library, Dublin*. Alexandra VA: Art Services International.

Z

Zeuner, F.E. 1952. "The Microlithic Industry of Gujarat", *Man* 52: 129–31.

*Zimmermann, Francis. 1982. *The Jungle and the Aroma of Meats: An Ecological Theme in Hindu Medicine*. Berkeley: University of California Press.

Page numbers in **bold** refer to captions

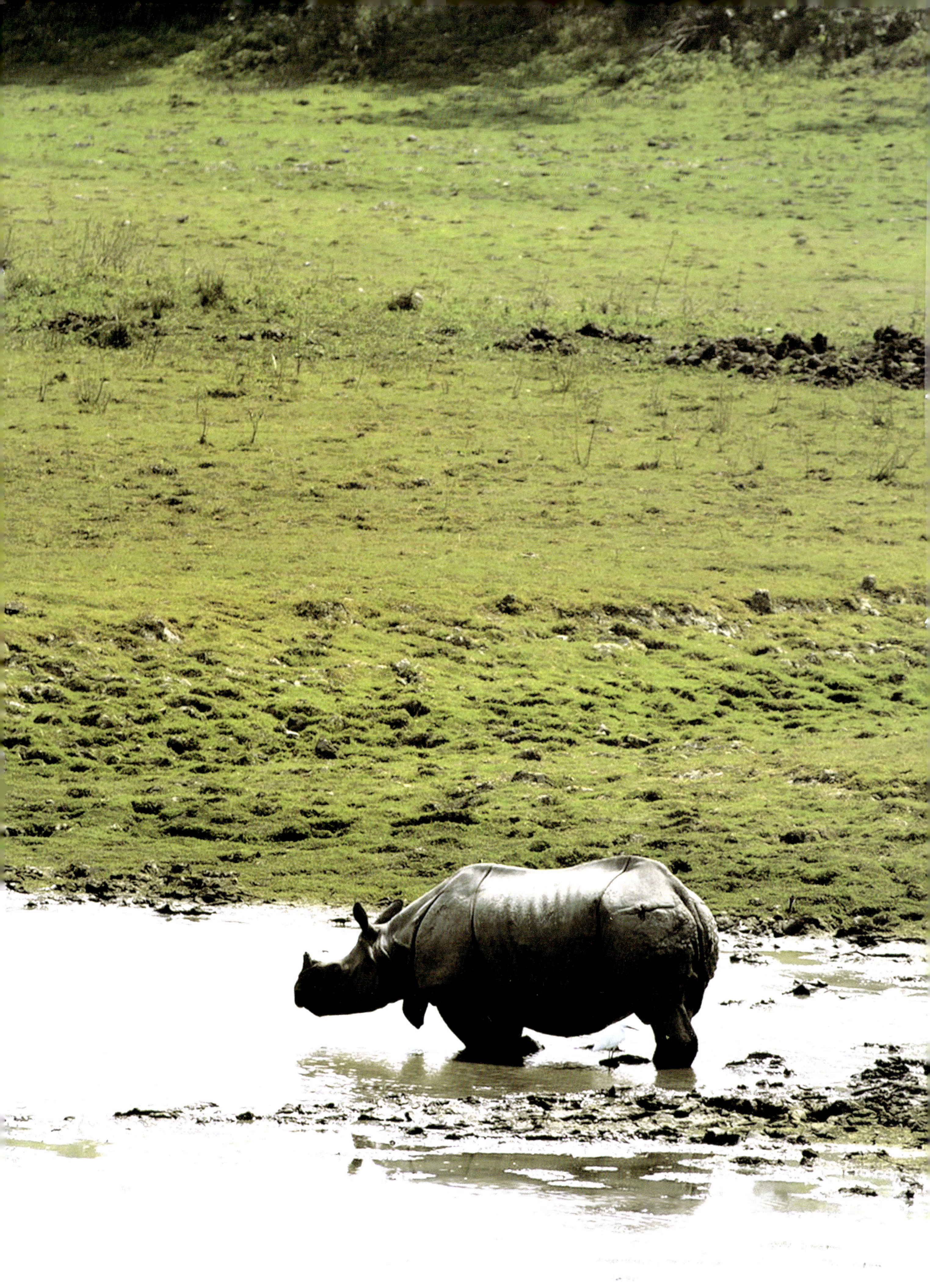

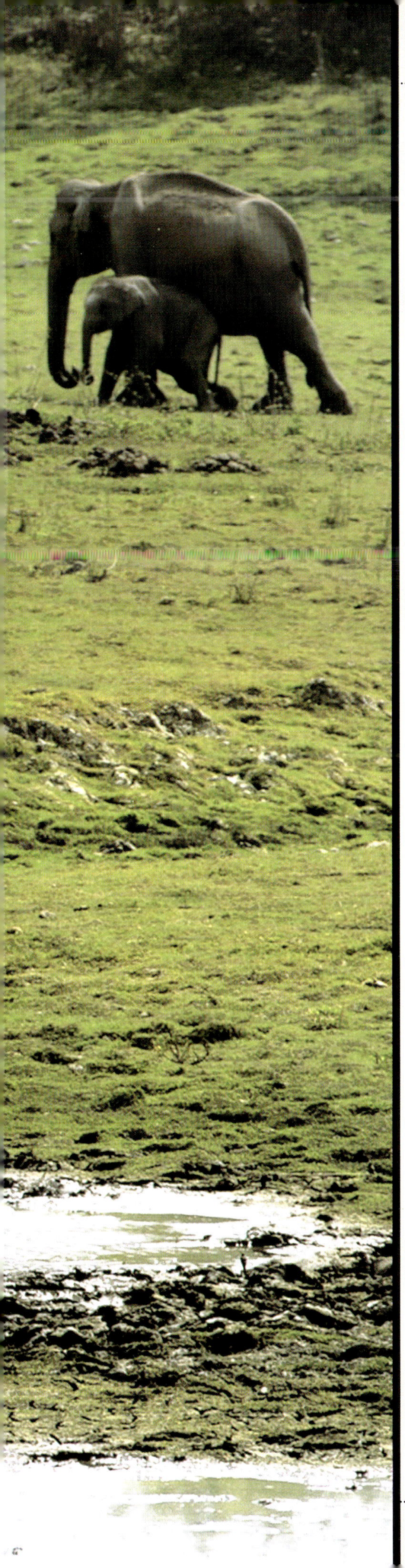

About the Authors

Divyabhanusinh is the author of *The End of a Trail: The Cheetah in India* and *The Story of Asia's Lions*. He is actively involved in conservation and was the president of WWF–India and Vice President of Bombay Natural History Society. He is a member of the Species Survival Commission, Cat Specialist Group of World Conservation Union (IUCN).

Asok Kumar Das is a senior art historian specializing in the study of Mughal art and culture. He has worked in and visited various museums and libraries around the world to examine their South Asian collections. Author of many papers, articles and books, he has recently been a Jawaharlal Nehru Fellow, a Tagore National Fellow for Cultural Research, and a Scholar-in-Residence at Jnana-Pravaha, Varanasi.

Shibani Bose is currently a visiting scholar in the Department of History, University of Minnesota. She received her MPhil and PhD from the Department of History, University of Delhi, and has taught at Miranda House, University of Delhi, and also at the University of St. Thomas, Minnesota. Her publications include essays in edited volumes, and her book *Mega Mammals in Ancient India: Elephants, Tigers, and Rhinos* is forthcoming.

▪ A rhino prepares to wallow, as a female elephant and calf pass by, Kaziranga. Usually they keep out of each other's path.
Photograph: Kakubhai Kothari.